Highway 17
The Road to Santa Cruz

by Richard A. Beal

Second Edition

The Pacific Group, Aptos, California

Cover photograph courtesy of San Jose *Mercury*. Photograph taken July 5, 1939 about 1.5 miles south of Los Gatos near today's Lexington Reservoir area. Old Santa Cruz Highway is on the left, new Highway 17 on the right. A several mile backup of homeward bound holiday cars illustrated the "bottleneck" problem where the new road met the last two miles of unimproved road.

SECOND EDITION

First edition: 14,000 printing

Book design and production by Barry Evans, Garden Court Press, 1001 Forest Avenue, Palo Alto, CA 94301.

Cover design by MasCom, 2608 Chanticleer Avenue, Santa Cruz, CA 95065.

Printed on recycled paper.

Publisher's Cataloging in Publication

Beal, Richard A.
    Highway 17: The Road to Santa Cruz / Richard A. Beal
    p. cm.
    Includes biographical sketches of early characters.
    Includes biographical references.
    ISBN 0-9629974-0-4

    1. Roads--Santa Cruz (Calif.)--History. 2. Roads--San Jose (Calif.)--History. 3. Roads--California--Design and construction--History. I. Title. II. Title: Highway 17: The Road to Santa Cruz.
HE356.C3   388.109   91-067400
               QBI91-1597

Dedication

To Kathy, Chandra and Andrew who provide the love that makes it all worthwhile.

Table of Contents

Chapter 3: Early Transportation Routes

Chapter 4: The Highway 17 Construction Project

Chapter 5: The Future of Highway 17

Chapter 6: Transportation Alternatives

Foreword

I commuted over Highway 17 for 16 years and lived to tell about it! That's over 4,200 round trips. Almost thirteen thousand hours or over 525 twenty-four hour days in the car, over 200,000 miles or 8 trips around the earth!

During my commutes I've driven Highway 17 with rain so heavy I couldn't see the front of my car, with snow falling at the Summit, been stuck in traffic for three hours without moving, had to detour for two and one-half hours each way because of the 1989 earthquake damage. I've seen hundreds of accidents, more than one death, avoided dozens of drunk drivers, been run off the highway by large trucks and watched helplessly as cars burned fiercely. I was rear-ended twice, once spun in a 360 degree circle on Big Moody curve avoiding another car, received three traffic tickets, listened to hundreds of hours of news, and dozens of times wondered about the history of the road and the towns along side it.

Both my father and grandparents drove the old Santa Cruz roads while on vacation. My father (age 75) remembers driving to Santa Cruz with his father in the 1920s. His aunt was killed during this period when his grandfather's Doret touring car went over the side of the road just north of the summit.

In 1990 I did research in the library and realized that no one had specifically ever focused on this "home away from home" for so many people in the South Bay area. And hence this book.

I care about making the information in this book as accurate and useful as possible. If you notice any errors or have suggestions for the next edition I'd love to hear from you. Send your comments to: The Pacific Group, Post Office Box 44, Aptos, Ca 95001, FAX (408) 662-0934.

Richard A. Beal

Introduction

Although the large increase in population has changed the area now transited by Highway 17, the largest single impact has been the roads.

Local historian William Wulf calls the building of roads "the single most important event in the history of the Santa Cruz Mountains." Roads have permitted area growth, commuter life styles, and routine 100 mile journeys for beach side picnics.

Highway 17 is the California State Highway connecting Santa Cruz and Santa Clara counties. Built in the 1930s, constant upgrading disguises its 52 years of age. Today more than 75,000 vehicles a day use the beautiful but dangerous road for recreation, access to work, vacations and commerce.

It rises from the Pacific Ocean to 1,800 feet in elevation and returns to almost sea level, spans 25 miles and offers some of the most beautiful scenery along any state road. But it is also a road that many people hate or fear. Accident rates have been very high over the years because far more people travel the road at higher speeds than the designers ever envisioned.

Most importantly, Highway 17 crosses an area rich in history. Early California pioneers settled in the Santa Cruz Mountains seeking privacy, land and wealth. The history of the early trails, the first toll roads, stage coaches, even railroads is largely unknown by most people using the highway. There are stories of greed, bigamy, fraud, speed, death, sexual crossovers, wealth and adventure. And lots of humor as well.

Besides the history of Highway 17 you will find current information on services, gas stations, emergency call boxes and telephones.

Read on, you'll be surprised! This is your history.

CHAPTER 1

Area History

Introduction

To understand Highway 17, we first need to begin with an understanding of the area and early history.

In the beginning was wilderness. High rugged mountains, abundant wildlife, giant redwood trees. And then humans began to arrive - first Indians, then Spaniards, Mexicans and finally Americans.

The new City of San Jose was on the main land route between San Francisco and Monterey, by way of Salinas. Agriculture, lumber and tanned cow hides emerged as commercial enterprises. Populations grew and it became profitable for merchants to find ways to connect San Jose and Santa Cruz directly.

Before there were roads for commerce and reliable travel, Santa Cruz County relied on ocean shipping for contact with the world. A road across the Santa Cruz Mountains was highly desirable. This is about how it all happened.

Early Regional History

The first humans came to the area 40,000 years ago - a diverse group of aboriginal descendants known collectively as California Indians. They originally came from Asia across the Bering Straight. These were not one politically organized body, but rather small groups of 200+ people with a common culture and similar languages.

1

The Santa Clara Valley's first inhabitants settled at least 1,000, and perhaps as much as 10,000, years before the Spanish arrived. The name "Costanoan" comes from the Spanish "costeños" meaning people of the coast. The word "Ohlone" is favored by descendants of the natives although anthropologists use the earlier term. [Payne, Howling, p. 11]

A large settlement of Indians had lived in the upper Sacramento Valley for centuries, and some migrated to the Santa Cruz Mountain area around 500 A.D. Originally the settlement was indigenous Indians from the Costanoan tribe of the Utian family.

These people lived in relatively small groups, building half sphere dwellings out of redwood tree branches and then covering them with branches and brush. They hunted and fished for food, with some gathering of local plants. Bow and arrow usage was common. Deer, bears, mountain lions, wildcats and various other animals were plentiful, and the climate mild because of the closeness to the ocean.

Although the Indian women wore aprons made of leather hides, the men went without clothing. Some remnants of the earliest inhabitants have been found near Lexington, including fire pottery, tools, fire cracked rocks and burned bones. About 5,000 of these sun worshippers were living in the Santa Clara Valley when the first white men appeared. Due to newly introduced diseases, a declining birth rate and exploitation by whites, by 1810 all traces of the original Indians have vanished.

The Spanish explorer Sebastian Vizcaino "found" the area now called the Santa Cruz Mountains in 1602. In 1769 Spain designated California as part of the Spanish Empire but the Spanish influence only reached 50-75 miles inland from the ocean and had little impact on the culture or the land. That same year the first Spanish explorers visited the Santa Clara Valley, and the Pueblo San Jose de Guadalupe was built in the vicinity of the current civic center. It was later moved to the San Jose downtown area. Mission Santa Clara De Asis was founded further north in 1777 by Father Junipero Serra, head of the Missions in Mexico. The first Santa Clara church was moved to various sites in 1779, 1784, 1818, and finally 1825. It continued to serve the religious needs of the community until 1851.

The Spanish called the mountain area Sierra Azul (blue mountains) and it was considered a rough undeveloped area, although in fact many Indians lived and hunted there. A Santa Cruz missionary wrote:

The adjacent mountains were wild and rugged, the canyons deep and dark with the shadows of the forest, coyotes broke the stillness with their dismal howls, and herds of deer slacked their thirst in the clear waters of the San Lorenzo. Grizzly bears were numerous, prowling about in herds, like hogs on a farm. [Young, p. 22]

Mission La Exaltation de la Santa Cruz was established in 1791 on the other side of the mountain range in the much more isolated area we now call Santa Cruz city. On his return to Mission Santa Clara, Father de Lasuen used a shorter route thus becoming the first non-Indian known to have crossed the mountains. The

Beautiful trees line much of Highway 17. *[Richard A. Beal]*

exact route wasn't recorded but most likely followed the basic route of today's Highway 17.

In the 1821-2 popular revolution, Mexico declared independence from Spain and the missions were abandoned. Wider settlement of California by Mexicans was encouraged, California Ranchos were offered free to any Mexican citizen willing to move and free trade quickly followed.

Monterey also had a mission and because of its location by the ocean, was considered a commerce center, along with San Francisco.

Raising cattle and cutting lumber were the primary activities in the Santa Cruz Mountains during the early 1800s. Later agriculture became the primary activity. Barbed wire was not invented until the 1870s and there was a tremendous need for wooden fences, posts and rails as well as flat lumber for buildings. The growing prosperity of the area attracted Americans and in 1846 the United States acquired California. Gold was discovered soon afterwards and the state population went from 15,000 to 100,000 in four years, then tripling again by 1860. As gold and large trees eventually diminished, the area turned to agriculture to support itself.

Geography and Animal Life

The highway was originally built to transit the mountains between Santa Cruz and Los Gatos - specifically the Santa Cruz Mountains which are part of the larger Coast Range. The highway covers considerable elevation change from 50 to 1,808 feet (552 meters) as it goes over the mountains which are about 25 miles wide. The road actually moves from North to South, although it is interesting that many local citizens think of it as an east-west road.

100 million years ago, the land under Highway 17 was located near Mexico City. It has slowly moved north, spent some time below sea level, and most recently (in geological terms) was pushed up by molten granite to form the current mountain features.

The mountain area of Highway 17 is covered with mixed needleleaf and broadleaf evergreen forests with some coastal redwoods. Maples, oaks, madrone, Douglas firs, eucalyptus, yellow acacia, bay and willow trees are especially beautiful in the Spring. The famous Coastal Redwood trees (Sequoia Sempervirens) are the tallest trees in the world and visible along many parts of 17. The mountains used to be covered with redwoods, but they were heavily cut in the late 1800s to handle the booming San Francisco economy - and again after 1906 when earthquake damage required massive rebuilding.

January mean temperatures at the summit run from 30-40 degrees, July from 70-80°. 90% of the precipitation occurs in winter with about 30 inches of rain average.

The San Andreas fault crosses Highway 17, along with at least 5 smaller faults, all on the Santa Clara County side between the summit and Lexington Reservoir.

Grizzly bears were common in the Santa Cruz Mountains until the mid-1800s when hunting the bears became a major activity - both for sport and because of their perceived dangerous nature. The last bear was reported killed in 1896, but as late as 1925 there were bear reports in the Pasatiempo area (Ayres, p. 6). Deer are still found in the area. Trout flourished in the streams until humans began to dam the water preventing their travel.

CHAPTER 2

Site History along Highway 17

Introduction

There is a tremendous amount of important history alongside Highway 17, as well as humorous folklore. This section gives the reader a brief walk through some of the areas most interesting history, before we proceed onto the early road development and later construction of Highway 17.

San Jose

There are no records of an Indian settlement at the site of today's city of San Jose, although Ohlone Indians probably traveled through the area. Even so, San Jose has the distinction of being the oldest city in Northern California, founded November 29, 1777 with Los Angeles following 4 years later. San Jose was the first government capital in California, with Governor Peter Hardeman Burnett presiding in 1849.

When the Spanish first invaded Northern California, it was primarily with the intentions of protecting shipping lanes along the coast and of locating valuable natural resources. These early Spaniards did not know how to farm and consequently ended up importing most of their food and supplies.

Overview Map

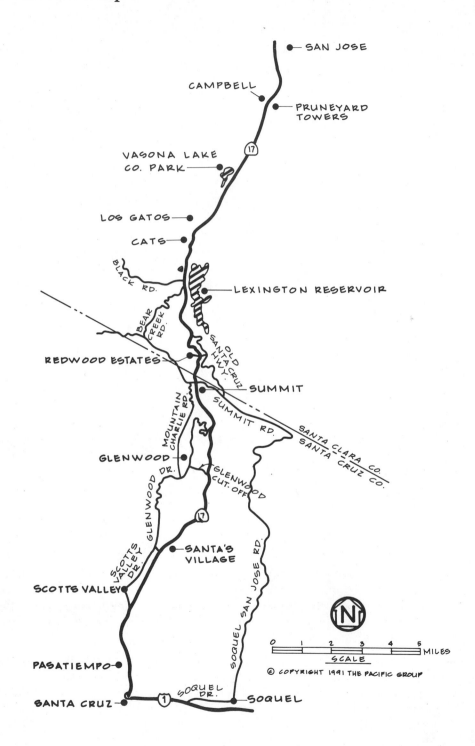

SAN JOSE

CAMPBELL

PRUNEYARD TOWERS

17

VASONA LAKE CO. PARK

LOS GATOS

CATS

BLACK RD.

LEXINGTON RESERVOIR

BEAR CREEK RD.

OLD SANTA CRUZ HWY.

REDWOOD ESTATES

SUMMIT

SUMMIT RD.

SANTA CLARA CO.
SANTA CRUZ CO.

MOUNTAIN CHARLIE RD.

GLENWOOD

GLENWOOD CUT-OFF

GLENWOOD DR.

17

SCOTTS VALLEY DR.

SANTA'S VILLAGE

SOQUEL SAN JOSE RD.

SCOTTS VALLEY

N

0 1 2 3 4 5 MILES
SCALE

© COPYRIGHT 1991 THE PACIFIC GROUP

PASATIEMPO

SANTA CRUZ

1

SOQUEL DR.

SOQUEL

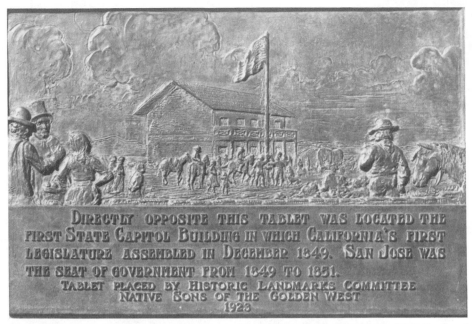

San Jose was the site of the first state Capitol building in 1849. *[UCSC Special Collections]*

Felipe de Neve (1727-1784) was a Spanish officer who quickly understood the need to develop local food production to sustain the European settlers. He scouted for level land near a major river that could be used for crop irrigation. The fertile area that is the site of today's downtown San Jose, along the Guadalupe River, was chosen for the first settlement. He then requested settlers from Mexico be recruited to farm the land, but impatient with the bureaucracy, decided to proceed on his own.

I resolved to withdraw nine soldiers with farming experience from the company of the presidio (of Monterey) and that of San Francisco, and signed up two settlers, along with three others who came here for that purpose. This made fourteen men. With their families the total number came to 66. With these I established the new pueblo of San Jose de Guadalupe on the 29th of November of (1777). The pueblo is located near the source of the Guadalupe River.... [Beilharz, p. 23]

Their first task was to build a dam across the river, so fields could be irrigated year around. Heavy winter rains broke the dam several years in a row, but eventually the settlers conquered it and by 1782 they were actually exporting grain from the city. These first settlers were Mexicans, Spaniards, Apache and local Ohlone Indians.

Although there was an active mission in Santa Clara (1777), in 1803-4 San Jose decided to build their own wood and adobe church dedicated to San Joseph and the Virgin of Guadalupe. The structure was damaged by earthquakes and in 1845 another church was built on today's site on Market Street (near the Fairmont Hotel).

This 1850 photograph is the earliest of Mission Santa Clara. *[Santa Clara University Collection]*

In 1822 Mexico took over the area from Spain and an oath of allegiance was administered to all the people of California in an attempt to guarantee loyalty. Mexico could not afford to continue to supply the area, however, and left the Californias to survive on their own. This only served to alienate California from the Mexican government. California's population was about 10,000 at this time. A few American explorers had visited the area and by 1841 an influx of American families were relocating to the area in hopes of claiming free land.

On May 13, 1846 the US declared war on Mexico and at least two small battles were fought near San Jose. The U.S. flag flew over San Jose by July of that same year.

In 1848 gold was discovered in California, only accelerating the population growth. By 1850 there were 4,000 residents in San Jose alone. Agriculture was the main industry with wheat, wine and a variety of fruits the most popular. Lumbering and some mining were also common businesses.

Today San Jose is the largest city in the Silicon Valley with a population over 800,000.

Campbell

In 1846, 250 prairie wagons from the mid west headed to California , including members of the Campbell family. The trip took about 6 months. Once they reached the Rockies, some of the wagons went north to Oregon while the rest took two routes to Northern California. Part of the latter group went through the pass later called Donner Pass and perished there when early snows caught them.

The Campbells and the rest of the group took a more southern route and arrived safely in Santa Clara. Census figures at the time showed only 800 Americans living in all of California. Mexico had declared war on the US while the Campbells were en route and the family soon became involved in helping defend the area with battles in Salinas and Santa Clara. In 1851 Benjamin and Mary Campbell purchased a quarter section of land on the site of the current town of Campbell.

By 1877 the S.P.C.R.R. had laid tracks through the land as part of the rail system connecting Alameda, San Jose, Los Gatos and Santa Cruz, and by 1885 Campbell had plowed a trail through the land and laid out the beginnings of a town. Merchants soon followed. The town was strongly prohibitionist and was one of the few in the surrounding area that had no drinking establishments.

Formal incorporation came in 1968. Today the town is primarily a suburb of San Jose but also hosts a considerable retail business community - including the PruneYard towers.

A complete history of Campbell, including lots of early photographs, can be found in the book *Campbell, the Orchard City* by Jeanette Watson, herself a native resident of Campbell.

PruneYard

One of the dominant features alongside Highway 17 is the PruneYard towers. Located east of the highway, between Hamilton and Camden, are two office towers and a nearby shopping center. In 1968 there were no buildings in Campbell higher than the old three-story cannery building. A young attorney named Fred Sahadi wanted to develop the area and construct a 18-story, 250-foot high office building. Campbell city officials were reluctant but soon agreed when they understood the amount of new tax revenues that would come to the city.

Sahadi's plan called for developing the Brynteson Ranch property into a 34-acre shopping center with two adjacent landmark high-rise buildings. The office building is still one of the highest structures between San Francisco and Los Angeles. Much of the structural steel had been prefabricated for an apartment building designated for San Francisco's Nob Hill, but the developer went bankrupt. Pittsburgh-Des Moines steel was caught with the materials and Sahadi was able to buy the steel at a discount price. By 1969 the tall tower (Tower

11

I) was finished and 3 1/2 years later the adjacent 10 story office building was completed as Tower II.

Although initially successful, Sahadi's company ran into financial troubles in recent times. Today the shopping center is owned by a new developer. The two towers have been placed in receivership by Home Federal, the primary mortgage company, and a new management company is responsible for day-to-day activities. Sebastian's Restaurant, which occupied the top of Tower I, is also no longer in business.

Vasona County Park appears in the upper left of this May 1988 aerial photograph looking north. *[California Department of Transportation]*

Vasona County Park

The popular Vasona Santa Clara County Park alongside Highway 17 is one of many Los Gatos public recreational facilities. The 400-acre property contains a large artificial lake, extensive picnic and grass areas, small rental sailboats and a very serene setting. White Peking and Mallard Ducks are prevalent in the park. Water for the Lake comes from the Lexington Reservoir. There is also a children's playground and the small scale Billy Jones Wildcat Railroad in the adjacent Oak Meadow park. The railroad engine is steam powered, and was originally built for a Venice, California land development company in 1903. Also popular is the carousel. It's one of 284 left in the United States, and features hand carved wood horses. Both the railroad and the carousel were purchased and restored by local volunteers.

Sacred Heart Novitiate

In 1886 the Jesuits established a center in the then remote foothills of Los Gatos. Today the center can be seen east of Highway 17, high on the hill across from Los Gatos. Officially called the Sacred Heart Novitiate, the center initially grew grapes and made sacramental wine but in 1892 they decided to enter the table grape industry.

Harvey Wilson had 39 acres of vineyard and 1,200 orange trees which were purchased by the Jesuits in March 1886 for $15,000. Three months later the first priests arrived and Father Pinasco was named the first superior. In the 1914-26 period, additional land was bought and a major building program started. They bought the Dr. Harry Tevis estate in 1934, adding about 250 acres of prime growing land. The old estate was renamed Alma College where priests-to-be received their last 4 years of 13 years of training.

The Jesuits retained 83 acres and the old winery which today houses the Mirassou Champagne Cellars. The Novitiate, one of the oldest wineries in California, was officially closed on January 5, 1986 and the center moved its operations to Berkeley. At its height it housed 120 students and 14 priest instructors. In 1984, after much debate, some of the land was sold to the Mid-Peninsula Open Space District for $3.7 million.

Los Gatos

In 1838, Mexican native Jose Hernandez of San Jose set out with his brother-in-law, Sebastial Peralta, to find a ranch site which could be homesteaded. They eventually build their adobe house in what is now Vasona Park, as part of a 6,631 acre 1840 grant from the Mexican government. The ranch was named after the Arroyo de Los Gatos, which flowed from the Cuesta de Los Gatos, or "Ridge of the Cats," so named after wildcats in the summit area. Today's town of Los Gatos is located within the original ranch site, near Los Gatos Creek.

James Alexander Forbes recognized that flour was being imported, while there were grain fields were nearby. All that was needed was a flour mill. He bought 2,000 acres of land and completed his mill in 1854 at a cost of $100,000. It ground 100 barrels of flour a day using water from the creek as the force to turn the grinding wheels. Forbes Mill eventually became the center for the town of Los Gatos. A 100 acre grant formed the basis for the original town, founded about 1864 and incorporated in 1887.

Later, in 1861, Los Gatos residents were excited when oil was discovered near Lexington, Alma and by Moody's Gulch. Many successful wells were drilled but the oil quickly ran out. Fruit trees were the next industry tried, this time with some success and land values increased rapidly.

Wildcats in the canyon area just south of Los Gatos had many unpleasant encounters with humans. The last known wildcat was killed in 1953 in the Redwood Estates area by Najor, a Cherokee Indian, for a bounty fee of $600.

Oil was discovered at Moody Gulch in 1861 and many wells were sunk. Dreams of quick riches quickly evaporated when it was learned that the quantity was low. [William A. Wulf Collection]

Santa Cruz Gap Turnpike Company Toll House. James Kennedy, who appears in this picture with his wife, built the toll house in 1867. It also served as their home. The house still stands in Los Gatos. *[Los Gatos Library]*

At 142 S. Santa Cruz Avenue you can see the oldest documented structure in Los Gatos. The "toll house" was the entry point to the road where James Kennedy collected tolls from 50¢ to $1.00. Originally the toll house was located at Lexington but was moved north to Los Gatos in 1867 . You can still see the original building at the corner of Santa Cruz Avenue and Wood Road although it has undergone various additions.

After the 20 year exclusive ownership of the toll road by the Santa Cruz Gap Turnpike Joint Stock Company had expired, the County was slow in taking over its right to public free access. Residents complained about having to still pay the toll and finally took matters into their own hands.

By January, teamsters, the turnpike's best customers, had become impatient with both the company and the county. They tore down the gate. When the company replaced it, they ripped it down again. D.B. Moody [the toll keeper], accompanied by two other officials of the company, was determined to keep charging tolls. The party, "well armed" according to a press report, went to the toll house and replaced the gate for the third time. About a dozen teamsters and their allies gathered to watch patiently as the new gate was installed. A reporter for the San Jose Weekly Mercury recorded what happened next. " Mr. Moody had no sooner taken his position at the breach, like Horatius of old, when the teamsters and their allies, seemingly oblivious of the warlike attitude of their adversaries, rushed forward, and, in less time than it takes to tell it, the gate was torn down and thrown over into the canyon. Mr. Moody held on firmly and came within an ace of following the gate over the bank......no weapons were drawn nor were any blows struck

15

Los Gatos as it appeared about 1890. The horizontal road across the picture was the early road to Santa Cruz (to the left). *[Los Gatos Library]*

on either side......As for the gate, it was fished out of the gulch, not without some difficulty." [San Jose Mercury, "When men were men and toll roads cost two bits," Bill Strobel, September 23, 1987]

During its 20 years of operation, the toll road enriched its stockholders $137,000!

Traffic on the toll road was very high, Historian William Wulf notes that on busy days there was a wagon every 15 minutes. In the mornings there was a two mile backup at the toll gate as drivers took empty wagons up into the mountains to get lumber that could be returned to San Jose.

Later Los Gatos was advertised as "Gem City" while Santa Cruz was "Surf City." The tourist dollar was highly desirable in the first half of the 1900s and the new Highway 17 was designed to connect these two growing cities.

The Cats

Just south of Los Gatos , near the Cats restaurant, are two large concrete cats standing near the entrance to a fenced off driveway.

"Leo" and "Leona" were created by the sculptor Robert Paine, and have guarded the entrance to an area known as Poet's Canyon since 1922. Writer Col. Charles Erskine Scott Wood and poet Sara Bard Field bought 34 acres in 1919 with the intention of building a "castle in the sky" - a place where they could write without being interrupted by uninvited visitors. They had a custom house built, planted a vineyard and installed giant wine casks at the bottom road as a protest to prohibition.

> I went up into my vineyard and lay down
>
> Upon the warm, red earth and smelled the sweet scent
>
> Of the little green flowers that tasselled the vines,
>
> From the top of the vineyard a cock-quail was crowing
>
> And over my head a white cloud was sleep in the sky.

Poems From The Ranch, Charles Erskine Scott Wood

Col. Wood was a very colorful figure - a West Point graduate, radical socialist, attorney, poet and philosopher. Wood met Sara while he was married to Nanny and soon fell in love with the beautiful poet - but his wife wouldn't give him a divorce. Finally, when Wood was 81, his first wife died and Sara and Scott were married soon after.

> Too long she has waited for me...waiting to be my wife
>
> And the end of the trail is not so far to go.
>
> We will not want more than the Indians do.
>
> Between the sweet-briar and the stream
>
> I will build a willow-wattled house, a chimneyed house for two,
>
> And a ditch to catch the sky blue gleam.

Poems From The Ranges, Charles Erskine Scott Wood

Sara wrote her best known poetry at the Cats, including "The Pale Woman," "Barrabas," and "A Darkling Plain" in 1940. The house was a mecca for artistic people and frequent visitors were artists such as violinist Yehudi Menuhin and writer Carl Sandburg.

The cat theme was picked simply because they liked the image, not because of the connection with the nearby town of Los Gatos. The sculptures are 8 feet tall, 10 feet in circumference and made of concrete because it cost less than bronze or marble. They were first modeled in clay. Paine lived at the Wood estate and accepted only a small fee for his year's work.

Immediately south of the Poet's Canyon entrance is the Cats Restaurant and Tavern (see Services for current information). Originally a stop on the old stage line, the Cats Roadhouse served as a way station for the horse-drawn lumber wagons on their way to San Jose, as well as a rowdy social club for area residents. At the time the road was first paved, around 1920, the Cats was one of the area's more notorious speakeasies and bordellos.

During the 40s and 50s the Cats building at different times housed a realty office, gun shop and sporting good store. The restaurant and tavern were re-established in 1967 and it is one of the few remaining roadhouses in the United States today.

The famous "cats" guarding the road to Poet's Canyon in 1925. The automobile is a Ford Model T (notice the spare inner tube for the tires). [*William A. Wulf Collection*]

Colonel (Ret.) Charles Erskine Scott Wood. Wood died at age 92 in his much loved"castle in the sky." [*Courtesy Horace Bristol*]

Lupin Naturist Club

One of the most famous businesses along Highway 17 is the Lupin Naturist Club owned by Glyn Stout. The Club is located at 20600 Aldercroft Heights Road, Los Gatos, phone (408) 353-2250. To get there turn at the Idylwild Drive exit and proceed one-half mile to the stop sign. Turn left (north) on Old Santa Cruz Highway and go a quarter mile to the fork in the road. Bear right onto Aldercroft Heights Road, follow it downhill across the white bridge. Stay right after the bridge and proceed about a block to the first driveway on the left. The small sign says "20600." Turn left and follow the pavement.

Introductory day visits are encouraged by offering a 50% discount from the regular day fee of $20. Membership plans are available and overnight stays are possible if pre-arranged.

The Lupin property was the site of a winery at the turn of the century, owned by a woman and her two daughters. Prohibition closed the winery and it was turned into a summer estate for a Southern Pacific official. In 1936 it became a naturist club owned by George Spray under the name Elysium Institute of California. Later names included the Rock Canyon Lodge and Lupin Lodge. In 1938 the members collectively bought the club and in 1946 George Bouffiel became owner. Stout leased the club in 1977 and exercised a purchase option in 1990.

Lupin occupies 110 acres, about 20 of which are developed with the rest left in their native state with access trails for hiking. They have 900 members. The 1989 earthquake damaged their main clubhouse and restaurant but new construction is in process. The facilities include two swimming pools, two spas, two tennis courts, sauna, volleyball, horseshoes, sunning decks, lawns and large shade trees.

Just so there are no surprises, Lupin is a "clothing optional" club where social nudity is a long valued choice of the membership.

Lexington

The town of Lexington, originally called Jones Mill, after the redwood sawmill in Santa Clara County, was founded about 1857 by Zachariah "Buffalo" Jones whose name allegedly came from his very loud voice. Jones was one of the first residents of the Santa Cruz Mountains. The original town site was covered when the Lexington Reservoir was built, along with the town of Alma and the northern most portion of the original Santa Cruz Highway.

The original toll house was located here but was moved in 1865 to Los Gatos. When lumber was a thriving business, Lexington was a significant town. About 3 miles south of Los Gatos, it had a hotel, livery and blacksmith shop, and eight sawmills. Jones bought a mill for $3,000 and called it Jones Mill in 1849. In 1857 he sold it and the new owner, John P. Henning, changed the name from Jones Mill to Lexington, after his home town of Lexington, Missouri. The population

Lexington in the 1860s. The Noviate vineyards are to the left. The dirt road is the toll road leading to the summit area. *[William A. Wulf Collection]*

By 1909 Lexington was a relatively large town. The Lexington Hotel can be seen facing the reader in the upper left corner. Notice there are two engines pulling the long passenger train. *[William A. Wulf Collection]*

was as high as 200. But as local lumber was depleted, the saw mills moved further up into the mountains. Los Gatos became more of a commercial center and Lexington went into decline.

Fishermen will be interested to know that the streams south of Los Gatos once boasted trout - the streams ran directly into the San Francisco Bay in those days. On the west side of the canyon near Lexington is a creek called Trout Gulch but the story is that it never contained any fish. Instead a Greek fish peddler lived up the canyon and only knew the English word "trout". He made regular trips to Santa Cruz and bought back a variety of fresh fish (other than trout) - and then on his trip home shouted out "trout" whenever he saw a potential buyer for his wares. Local residents soon began calling him Trout and hence the name for the Creek.

When Lexington Church installed its first organ, Jones is reputed to have told the preacher, "You've got the devil in our church now. Step on its tail and hear it squeal."

Lexington Reservoir

The Lexington Reservoir is owned by the Santa Clara Valley Water Conservation District. The Santa Clara County Parks and Recreation Department manages the recreational uses of the reservoir under a lease with the SCVWCD which expires in 1995.

The reservoir is part of a larger County program to back up water in "percolation" ponds that drain water from the mountains into the underground water table, rather than just running off. Originally the mountain streams flowed all the way to the San Francisco Bay. Santa Clara Valley's dam system was designed by Frederick H. Tibbett and Stephen E. Kiffer with Walter Hunt, Chief Engineer in charge of construction.

Highway 17 was re-routed in December 1951 and the dam construction started in the Spring of 1952, finishing that fall. It was dedicated in April of 1956 by Governor Goodwin Knight, with the ceremony held at the relocated Alma Fire Station (the station was moved to its present location in 1955).

The dam was built across the Los Gatos Creek at a cost of $5 million. It took 34 weeks to construct the 170 foot high dam which is 1,000 feet thick at its base and 1,000 feet long. There is a 150 foot concrete spillway at the west end and when full the Reservoir holds 25,100 acre feet of water. That's enough water to supply 25,000 families with water for a year.

The resulting Reservoir covers the north most point of what was originally the old Santa Cruz Highway connecting Los Gatos and the Summit area along with the former towns of Alma and Lexington. At the time the dam was proposed, fewer than a dozen families lived at Lexington and about 50 families at Alma. All were relocated and the old road was abandoned.

Aerial shot looking north from above Redwood Estates in May 1988. Lexington Reservoir was almost drained and the old road can be seen as the straight line through the Reservoir. Some of the concrete culvert and roadwork still survives today and can be seen when the water level is low. *[California Department of Transportation]*

This view looking east across Lexington Reservoir shows a barely visible portion of the original Santa Cruz Highway (center of the picture) later cover by the water. This picture was taken the summer of 1991 when the water level was quite low. *[Richard A. Beal]*

The wooden flume shown near Alma brought water to Los Gatos to help power Forbes Mill and later to generate electricity. *[William A. Wulf Collection]*

The San Jose Water Company Montevina Treatment Plant facility at the north west corner of the Reservoir was built in 1960.

Before the Reservoir, a 3-foot wide wooden flume ran high on the east side of the Los Gatos canyon, bringing water from the mountains down to the town of Los Gatos. Originally the flume was to bring water to Forbes Mill to turn the mill stones. Later the starting point of the flume was moved to Lexington and then higher in the mountains to gain more of an elevation drop so that the water force was greater. Eventually a 200' drop was realized. This water also was used to generate the first electricity for the town of Los Gatos.

Recreational uses at the reservoir are secondary to the flood control. Because of the low water levels, the Lexington Reservoir Park has been closed since late 1988. There is a small boat launching area (power boats are prohibited). The Los Gatos and the University of Santa Clara rowing teams practice when there is enough water in the lake and have a boat storage building on the property. Currently there are no fish in the water. There is also a walking trail - the Los Gatos Creek Trail - between Los Gatos and the Reservoir. The Park was very popular when it first opened with 25,000 people using it one week in July of 1984.

Every few years there is discussion about expanding the recreational facilities at the lake, but local residents strongly oppose this.

Alma

About a mile south of Lexington, Alma boasted a stage coach station, hotel, store and six lumber mills. Founded by Lysander Collins in the 1860s, Collins took lumber instead of wages from the Howe mill where he worked and used the lumber to create the first buildings. The Forest House hotel was a very large 2-

This color postcard shows an early street scene in front of the Alma General Merchandise Store about 1910. *[Covello and Covello Photography]*

story building complete with saloon. Later there were land ownership disputes (very common in those days) and the village became Alma by edict from the US Post Office. Alma is Spanish for "soul" and was regarded as the "soul" of the redwood area. Incidentally, the same postal inspector also is responsible for the name "Patchen."

As Alma prospered, Lexington declined. By 1905 Alma was only a flag stop on the railroad - meaning that the train only stopped if someone at the station waved a flag indicating there was freight or a passenger to pick up. The town was eventually covered by the Lexington Reservoir project. A few buildings were moved but the original site is completely gone.

In 1906 Dr. Harry L. Tevis, a retired physician from San Francisco, purchased land in the area from James C. Flood. At one time the land was known as Fish Ranch. Dr. Tevis spent $750,000 on buildings and the grounds. He grew rare flowers and shrubs collected from all parts of the world here. At the height of its fame, the Tevis gardens boasted prize winning giant dahlias, lilies, roses, fuchsias and nandia shrubs. When Tevis died Yehundi Menuhin, the violinist, wanted the buy the home but it went instead to the Jesuit order who began Alma College.

The Highway 17 routing went right through these prized gardens. Gardeners from the University of Santa Clara, aided by California Conservation Corp (CCC) crews, removed dozens of rare shrubs and trees from the once famous Tevis gardens near Alma and took them to the campus for replanting. Father G.A. Gilbert of Alma College spearheaded the effort.

Holy City

The story of Holy City is one of the most colorful along Highway 17 and I highly recommend an excellent unpublished paper by Joan B. Barriga called "The Holy City Sideshow." Author Betty Lewis is also planning a book to be published in late 1991 on the subject.

Holy City was established by "Father" William E. Riker, known to his band as "The Comforter," along with 30 members of his Brotherhood of The Perfect Christian Divine Way in 1919. Holy City was believed by its devout followers to be the future center of the world, with Riker and his wife as King and Queen.

William Riker was born in 1873 in California. As a young man he was good looking and had a magnetic personality that worked especially well on women. He started as "Professor" Riker doing palm reading and then went on a national tour doing a mind reading act. While in San Francisco, the District Attorney filed bigamy charges against him and Riker quickly fled to Canada leaving both his wives behind.

In Canada he developed a new philosophy called The Perfect Christian Divine Way that emphasized white supremacy, total segregation of the races, no

25

Holy City in the 1920s (looking northeast). *[Betty Lewis Collection]*

drinking of alcohol, separation of the sexes and being "born again." He returned
to San Francisco and set up a commune, requiring his followers to give their
money to him to free themselves from worldly concerns. In 1918 he bought 75
acres of land in the mountains about 10 miles south of Los Gatos for $6,000 or
$7,000, later expanding it to 200 acres. By this time his congregation was up to
30 people and Riker had them building and running tourist oriented businesses.

Riker was apparently exempt from his religious rules of being separated from
women because he soon married Lucile. One of his earliest women disciples,
Frieda Schwartz, became irate at this and filed suit to recover her funds. There
was great publicity over the suit but it only served to draw curiosity seekers to
the area. Soon Riker had a $100,000 a year tourist trade business going alongside
the Old Santa Cruz Highway including a restaurant, comfort station, gas station
and observatory where for 10¢ visitors could see the moon through a telescope.
Billboard signs enticed visitors with: "Holy City answers all questions and
solves all problems" and "See us if you are contemplating marriage, suicide or
divorce." Although Holy City was advertised as a religious place, in fact no
church was ever built.

In 1926 Holy City was incorporated with all property and income in Riker's
name. In 1929 he established radio station KFQU [Adult readers might want to
try pronouncing the radio station call letters silently to themselves]. This was
the second licensed station in California, although Riker's license was revoked
two years later for "irregularities". While it was transmitting it offered a broad
variety of popular programming including a one-half hour show with a Swiss
yodeler. In the 30s Holy City's population grew to 300, mostly out-of-work
drifters from the depression who could find work there.

In 1938 Riker ran as a minor candidate for California's Governor (he later ran three more times) and in 1942 he was arrested by the FBI for pro-German sentiments. This was during World War II and he was writing support letters to Hitler. The famous attorney Melvin Belli represented Riker in court and freed his client - but then Riker filed suit against Belli, claiming the lawyer had defamed his character during the trial by repeatedly referring to him as a "crackpot". Belli also won that case and got his $7,500 fee.

About this same time, Highway 17 opened and there was a dramatic drop in traffic along the Old Santa Cruz Highway running through Holy City. The city began to decline, and in 1959 a complicated real estate transaction threatened to take control of the property away from Riker, then 86, who fought it and lost. A series of mysteriously set fires then destroyed most of the buildings. Finally in 1966, at age 93, Riker shocked his few remaining disciples by announcing that he had converted to Catholicism. He died soon after in 1969 at Agnew State Hospital.

None of the original buildings remain, with the exception of Father Riker's Victorian-style house. The Holy City Art Glass building erected in the early 1960s is on the original site.

Holy City in the early 1930s. [*William A. Wulf Collection*]

Redwood Estates

Redwood Estates was a commercial land development aimed at San Franciscians who now had automobiles and longed for a country place to visit on weekends. Two hours away from the city, nestled among giant redwoods, many visitors stopped at the sales office on their way to and from Santa Cruz and came away as land owners. Harry W. Grassle was the manager of the development built in the mid-1920s and boasted:

It is on a paved state highway and to get there you have no ferries to cross. It is immediately accessible from either side of the Bay. On the San Francisco side the highways are being rapidly improved...The Bayshore Road, 125 feet wide, is now assured. The Skyline Boulevard is now completed as far as Redwood City....The State Highway Commission is going to double the width of the road from Los Gatos to Santa Cruz so we are assured of a four track highway passing by our gates. [Cabinland Magazine, February 1927]

Various size view lots ranging from twenty-five feet square to one acre were offered at prices from $100 to $1,000 and the developers had various cabin building plans available to interested residents. This area, at one time known as the Mountain Springs Ranch, offered free water, wiring for electricity, a large community center complete with a swimming pool and, for automobile owners, a "passable" road to each lot.

Before 1926 there was a large wooden bear holding an umbrella south of the entrance, and in 1927 the Redwood Estate company built a 15 foot high Dutch style windmill building at the turnoff from Highway 17 to attract attention.

The Redwood Estates area still exists along Highway 17, although today it houses mostly permanent residents.

Gail Lloyd, Hollywood movie star, helps advertise mountain cabins at Redwood Estates. *[Cabinland Magazine]*

Electricity - Graveled Roads - More Reservoirs
Swimming Pool!

*(As for the pictures on this page, and when and why and how they got that way—
please see the preceding page!)*

Every cabin on the Redwood Estates may now boast of electricity's comforts! Owners desiring electric service should have their cabins wired and then put in the usual application with the P. G. & E. at San Jose.

EVERY MODERN IMPROVEMENT

The home-comforts of electricity comprise one more of the forward strides we have been contemplating. Our one consistent plan is to give

every modern convenience and comfort in the midst of this mountain wonderland, so that your days or weeks of rest and recuperation in your cabin will really *mean* rest and recuperation.

A TELEPHONE IN YOUR CABIN!

At very reasonable rates, the Los Gatos Telephone Company will install a telephone in your cabin — you have but to make application at their Los Gatos office.

ALL ROADS GRAVELED

In making every cabin accessible by auto all the year around, the final convenience of graveled and oiled roads is now also an accomplished work. All completed roads are now oiled, graveled, and ready for your constant use.

TWO MORE RESERVOIRS

No other property has seen such rapid development. Not content with our first 125,000-gallon reservoir—full to the brim with the incomparably pure and healthful spring water of the Redwood Estates—the contract has been let for two more reservoirs and work is starting at

once. These additional reservoirs will be of the same type of construction as the one already completed and in use. In this way, every cabin on the entire Estates is assured of adequate and constant water-supply—piped free and supplied free to every cabin—always!

SWIMMING POOL

The contract has also been let for the building of the swimming pool. It will be completed within 45 working days, and will be freely available for the enjoyment of those who own lots on the Redwood Estates.

OWNER'S DAY

"Owner's Day", Sunday, May 1st, was a gratifying success—over 500 of the Redwood Estates family gathered on the property and spent the delightfully clear and sunny day in games and hikes.

1000 LOTS NOW SOLD

The thousandth lot has been sold! Impartial observers who visit the Redwood Estates—observers from Florida, New York, and all thru

the country—also from Los Angeles—are unanimous in their prediction that, with the marvelous scenic and climatic beauty united with modern conveniences, every lot on the Redwood Estates will be sold out *within the next few months!*

You who have been planning and promising yourself and the family one of these paradise *garden-spots*—now is the time to ACT!

All the modern conveniences: electricity, gravelled roads and a swimming pool! *[Cabinland Magazine]*

Patchen

Patchen Pass is the actual name of the spot where Highway 17 crosses the Santa Cruz-Santa Clara County line. Most people refer to it as "the summit."

In the late 1970s there was considerable controversy about the name - which was decided by Los Gatos Councilman, later Mayor, Albert Smith. Local historians protested that the area had been designated as "Cuesta de Los Gatos" by the explorer Fremont. Cuesta de Los Gatos means "wildcat ridge" in Spanish. Smith ignored the advice and even though the ridge was outside of the Los Gatos city

Redwood Estates entrance. The Dutch windmill was built to catch the attention of passing motorists and attract them to the sales office located immediately behind the windmill. *[Bruce Kennedy Collection]*

limits, he had signs made and installed them at the summit. Caltrans later removed them. Politicians, Caltrans, the U.S. Department of Interior and local newspapers had a field day arguing about the name for the next three years before Patchen Pass was officially designated as the official name.

There was no official town called Patchen. The name refers to a 1872 Post Office and stage stop, about a mile north of the Summit on the Santa Cruz Turnpike at the Mountain Charlie Road fork. The post office was situated on the Fowler ranch, on the west side of the old highway at the junction. The building burned down in the 1950s. Across the highway was the stagecoach barn, which was torn down in 1949.

As the story goes, the [Post Office] inspector stepped out of the stage coach in front of Fowler's place to encounter a man sitting on the doorstep busily sewing. 'What are you doing?' asked the inspector, seeking to establish friendly relations with the natives. 'Patchin', replied the old man, and that was that. [Clark, p.252].

Another less colorful version stated that the town was named after the famous race horse of the 1850s, George Patchen, but the Los Gatos Times Observer disputes that story in a 1978 article. In any case the location of the post office was moved several times over the next 50 years.

There is a historical marker at the site of the former Post Office, at the intersection of the old Santa Cruz Highway and Mountain Charlie Road.

Young records that:

31

Patchen in 1866. The dirt road in the background was the old toll road. *[William A. Wulf Collection]*

Previous to the establishment of the post office, mail had been deposited in a hollow tree nearby and the customers all sorted their own mail. [Young, p. 29]

German cabinetmaker John Martin Schultheis (known as "Mart") and his wife Susan Byerly homesteaded land about a mile from the Post Office in 1852 and built the first permanent home in the mountains - a log cabin. The Schultheis house is believed to be the oldest building still standing in the Santa Cruz Mountains and is located at the rear of 22849 Summit Road. The Schultheis family bought more land and eventually owned 75 acres, mostly used for growing prunes. Susan was a well known nurse and delivered many of the babies born in the area.

By 1885 Patchen consisted of a post office, store and a few hotels serving local summit residents and travelers along the Mt. Charlie road. An Episcopal Church was built nearby in 1899 as a branch of the Los Gatos Church. In 1887 the Schultheis family donated land for the Summit Opera House. The 40' x 75' Opera House was erected with volunteer labor in 1890 and used for community dances, classical music (Mart Schultheis was Director of the orchestra) and of course recitals by opera singers.

Another prominent citizen in the Summit area was Lyman John Burrell, who arrived in 1851 from Ohio. He was married to Clarisa Wright and in 1856 planted an orchard and vineyard. Soon many other settlers followed his lead and raising fruit became a major industry. The Burrell land was right on the county line about 1.25 miles south of Wrights and at its height had a store, blacksmith,

32

Presbyterian Church, and several hotels. There is still a Burrell School on Summit Road.

Wright's

Wright's existed because of the railroad. Situated near the summit, the town was located at one end of a mile long tunnel (later designated as #2) that had its other end near Laurel. Before the railroad construction, only a shack marked the spot of the future town. But soon a notorious saloon sprang up in the otherwise quiet mountains, featuring a drink called the "discovery" which called for diluting one gallon of raw whiskey with four tablespoons of water - and drinking it all at one sitting! After the construction crews moved on, the saloon closed and the town became more respectable, although life in the Santa Cruz mountains was still quite wild.

The town was named after the Reverend James Richard Wright, a retired Presbyterian minister from Michigan, and his wife Sara Vincent who settled in 1869 on 48 acres they had been given as repayment of a debt. They had 10 children. Wright built a summer resort and when the train came in 1879 he also built a railroad station and telegraph office.

Soon other businesses followed, including fruit packing houses such as the Earl Fruit Company and the Pioneer Fruit Company which packed the local fruit included grapes, pears, apples, cherries and plums.

Wright's Station Hotel. [UCSC Special Collections]

Railroad employees pose for this picture at the Wright's tunnel as Engine 17 heads north from Laurel. Notice the water spillway in the background. *[William A. Wulf Collection]*

Farmers bringing their fresh fruit to the Wright's Station. Middlemen purchased the produce and shipped it throughout the United States. *[San Jose Historical Museum]*

Wright's hotel, saloon and blacksmith shop about 1885. The businesses later moved to the other side of the train tracks when these buildings burned down. *[William A. Wulf Collection]*

Wright's with the train fully operational. The tunnel leads to Laurel. *[William A. Wulf Collection]*

Wright's Grangers stop to pose for a picture on the way to Skyland to found a new Grange at the Adams Ranch (1880). *[San Jose Historical Museum]*

Summit resident Jeanette (King) Andrus recounts:

In the old days, the railroad would notify my father when refrigerator cars would be on the siding at Wrights. He and the other local ranchers would then take their fruit and produce down and load the cars. Usually the cars would be picked up by the night freight from Felton. Sometimes the fruit would go all the way to New York. I knew this because when I was a little girl I used to help fill fruit crates. On the bottom of some of the crates I would leave a letter telling who I was and from where the fruit was coming. I received many replies to my letters, some from as far away as New York. I once even received a silver bracelet from a man who assumed I must have been a rich California farm girl. When father found out, he put a stop to my letter writing. [Hammond, p. 104].

By 1882 Wright's was the major shipping center in the mountains. But on the fourth of July 1885 the whole town burned down when an overheated stove set fire to the hotel. A. J. Rich acquired the property and rebuilt the town on the east side of the tracks. Wright died in 1896, his wife in 1908.

Early advertisements touted deer hunting, trout fishing, a swimming pool, and tennis only a 2 hour train trip from San Francisco (cost $2.50 round trip). The "picnic train" had a stop here and tourists often got off to spend the day in this beautiful area. The railroad built a picnic area named Sunset Park and it soon became an extremely popular place for locals to meet.

In 1936 the San Jose Water Works acquired the land to protect its watershed and the town was abandoned after 1940 when the trains stopped running. No buildings remain today.

Idylwild

Louis Hebard was the first white settler of the present Idylwild property, located east of Redwood Estates. He arrived from New York in 1857 and homesteaded 160 acres. Hebard was well known in the area for his carpentry skills and worked on many homes, schools and mills. In 1863, his first wife, Lucinda, died and in 1875 he married Lodiska Ann Girard. There is a Hebard Road named after him, just past Idylwild Road.

Highland and Skyland

A man named Dodge leased land in 1867 to establish a vineyard and called it Highland Hill. It soon became a noted fruit district and in the area local residents today refer to it as Skyland. In the late 1890s it boasted a post office, general store, the Highland School and even a newspaper. On November 11, 1887 four mountain residents gathered in the school house to form the Hyland Presbyterian Church. They met in homes until the church was completed in 1891.

As the area declined, the buildings were abandoned and even the church seemed doomed. But in 1949 ranchers Henry and Ruth Von der Mehden saved the building by paying off the original loan and organizing the "Harvest Festival" as a fund raiser to help restore the building. In 1951 the church became non-denominational, later affiliating with the United Church of Christ, and was renamed the Skyland Community Church. Today you can attend the church at 25100 Skyland Road (Los Gatos mailing address), phone is (408) 353-1310.

There is no longer an official Skyland area on the maps, although many local residents still refer to themselves as Skyland residents and the active church boasts 130 members.

Orchards were extremely popular in the Santa Cruz Mountains. This picture was taken in winter near Wright's. Date unknown. *[UCSC Special Collections]*

Loma Prieta

Loma Prieta is the highest peak in the Santa Cruz Mountains (3,791 feet) and can be seen to the east of Highway 17 from several places. In cold weather it often has a small amount of snowfall. Today it houses many antennas for various electronic transmissions but is best known for its beautiful views of both Santa Cruz and Santa Clara counties. Loma Prieta means the "dark mountain."

Because of the height dominance of the mountain over the rest of the range, we commonly hear of the "Loma Prieta earthquake" or the "Loma Prieta fire" even though the peak is only part of the much larger mountain range.

Summit

Summit Road follows the crest of the Santa Cruz Mountains basically along the dividing line between Santa Clara and Santa Cruz Counties. It connects with Skyline Boulevard which starts in San Francisco and changes its name to Summit Road at Bear Creek Road in Santa Cruz County. The road was originally built around 1916. Two miles east of the summit (Patchen Pass) is the Woodwardia corner where the Old Santa Cruz Highway crosses the summit.

Before the 1906 earthquake, there were several good size lakes near the summit that people used for fishing and recreation but the earthquake disrupted their water flow and they dried up.

This Woodwardia business offered a restaurant, food, cool drinks and dancing. A sign in the background announces, "we serve chicken on Sundays." *[William A. Wulf Collection]*

To get supplies, mountain residents went to either Santa Cruz or San Jose on regular visits. A few merchants ventured directly into the hills. Mardi Bennet writing in the San Jose Mercury records:

Eleanor Robison, who now lives in Willow Glen, remembers her grandfather Charles Roberts driving a canvas-covered wagon from Pepper and Gelatt's Butcher Shop in Los Gatos up to Skyline Blvd. twice a week to deliver meat. Roberts carried a gun because once a month he collected on accounts. On one occasion he exchanged shots with a highwayman - who got away. Roberts was left with a bullet hole through the canvas, and an old pair of overalls near the road from which the would-be robber had cut a denim mask to hide his face. [San Jose Mercury, August 9, 1989, p.11.]

The businesses at the summit didn't come into existence until the mid 1950s. Charles C. Martin originally owned property from the summit to Glenwood and over the years portions were sold off while others remained in the family. Richard Lisle Shore owned property just south of the Summit when Highway 17 went right through it in the late 1930s. Dick ran an ice and storage locker business in Los Gatos but decided to try his hand at the restaurant business in 1954. He cleared land east of the highway and built the first summit restaurant, a hamburger stand, calling it Cloud 9. Since then the restaurant has changed hands several times and is now called the Mountain Top Restaurant, owned by Tony Hwang.

Across the street, the present Summit Inn building was built two years later in 1956 by Mr. Hoeffler where he operated a restaurant for years under his name. Today it is the Summit Garden Inn owned by Jimmy Pang.

The Summit Properties building, now located just north of the Summit Inn, has actually been moved 3 times. The original building was a 2-story ranch house that was moved several hundred feet west to the present location in 1936-7 because of the Highway 17 construction. That building was destroyed in the 1989 earthquake. The current office is a re-build that was completed in 1990.

There was also a small refreshment stand on the northern end of Inspiration Point, just south of the summit restaurants, but it was demolished as part of the Highway 17 realignment in the 1930s.

Over the years there have been several unsuccessful attempts to add businesses to the Summit area such as an emergency medical center, hotel, gas station and convenience store. The two restaurants and real estate office are all located within Santa Cruz County and the counties official policy is:

[To] allow the limited expansion only of the existing restaurant uses on Highway 17 at the Summit and only if mitigations can be found and implemented for existing and potential traffic impacts.

The County is primarily concerned with the potential traffic hazard if more cars used the area, and the environmental impact on an area that has severe water problems, earthquake faults, etc. A gasoline station tank leak, for example, could accidentally contaminate the entire water supply for area residents. Local summit people share these concerns but more importantly want to preserve the rural atmosphere in the area.

Mountain Charlie Road

Just west of Highway 17, off Summit Road, Mountain Charlie Road was once part of the McKiernan Toll Road from Scotts Valley to the summit. What remains of that original road, built in 1868, is now officially called Mountain Charlie Road. The next chapter will have more information about the early mountain roads, but one famous mountain personality needs to be introduced now.

Charles Henry "Mountain Charlie" McKiernan was the second white man to live in the Summit area (settling in 1851). The first was a hunter called Daniel Post. Born in Ireland in 1830, Charles travelled to Australia while in the Army.

This large painting of Mountain Charlie McKiernan hangs at the Forbes Mill site historical museum in Los Gatos. [*San Jose Historical Museum*]

When his enlistment was up he headed for the California gold strike. After a year in the mines, he came to the Santa Cruz Mountains and homesteaded a home, initially living alone. He hunted, tried raising beef and even did some gold mining without luck.

McKiernan is most famous for his losing fight with a grizzly bear. In 1854 while hunting with a friend, he was surprised by a 1,000 pound mother bear with two cubs. Mountain Charlie was seized by the bear, which crushed the front of his skull. The other hunter eventually managed to distract the bear. McKiernan recovered but had a metal plate, made from two Mexican dollar coins, temporarily fitted into his skull. At age 26 he married Barbara Berricke Kelly, an Irish nurse who had nursed him back to health after the nearly fatal grizzly bear attack. Barbara bore him 7 children.

There are many other bear stories in the Santa Cruz Mountains, but by the late 1800s most of the bears had been killed.

In the 1870s McKiernan started a stage coach business and later became one of the most successful businessmen in the area. McKiernan's cabin near the summit was often a stopping spot and became known as Halfway House. Barbara cooked meals for the stage coach passengers while Charlie helped change horses on the wagons. After the new railroad diverted the toll road's business, Charles and Barbara moved to San Jose in 1884 where Charlie eventually died in 1892

Life in the Santa Cruz Mountains in the late 1800s was very wild. Small time bandits used the area as hideouts - picking on travellers as well as making forays into the "big" towns. Local historian James Addicott records:

McIntyre, who raised cattle on the Zayante Creek Flats, was murdered by two men frenzied by drink who went after McIntyre's hidden treasure [money]. He was mercilessly butchered and his body burned in his pioneer mountain cabin....but they were chased and caught on Mt. Charley road by a San Francisco posse who hung them on the little old Los Gatos wooden bridge on Main Street. [Sentinel, December 17, 1950]

In 1982 a plaque was installed at the northern most end of Mountain Charlie Road (near the Summit Road Overpass) stating:

Mountain Charlie Road. In 1858 the Santa Cruz Turnpike Company issued a contract in the amount of $6,000 to Charles Henry "Mountain Charlie" McKiernan and Hiram Scott for the construction of a road. The road from the Scott House, located in what is now Scotts Valley, to the Summit was later known as the McKiernan Toll Road. It subsequently became part of the Santa Cruz County road system on August 27, 1878 when the Santa Cruz Board of Supervisors issued a warrant for $600 to Charles McKiernan in consideration of such abandonment by him. Dedicated October 9, 1982. Mountain Charlies Chapter No. 1850, E Calmpus Vitros, "Right Wrongs Nobody"

All that is left of Mountain Charlie Road is a 5.2 mile section of a beautiful narrow road that goes from the Summit to Glenwood Highway, and a 2 mile section that goes north from the Summit to the old Santa Cruz Highway.

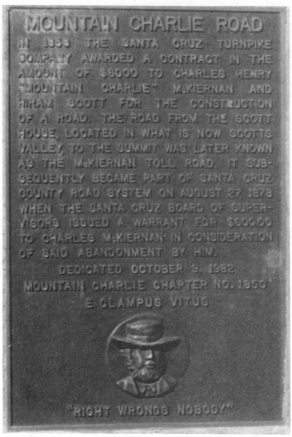

A monument to Mountain Charlie was erected by friends in 1982. *[Richard A. Beal]*

Laurel

Laurel was an "instant" town that sprang up when the railroad construction pushed through Tunnel 2 around 1880. The town was in an open spot between that Tunnel and Tunnel 3 that continued south towards Glenwood. The Laurel site is 3 miles north of Highway 17 on Laurel Road at the intersection with Schulteis Road and Redwood Lodge Road.

Today there is nothing left of the town except a commemorative marker:

Laurel. Once thriving railroad and sawmill town, known as Hyland when the first through train ran in 1880. Name Laurel adopted about 1885. F.A. Hihn built sawmill here in 1889. Laurel Mill supplied much of the lumber to rebuild San Francisco in 1906. Little remains today except tunnel portals and memories. This historical marker placed by the County of Santa Cruz and the Santa Cruz Bicentennial Commission October 1969.

After the railroad construction passed, it became the headquarters of the Frederick A. Hihn timber holdings and had a post office from 1882 to 1953.

The first Laurel Mill (1900-1906). *[UCSC Special Collections]*

J.C. Laurel grocery store and gas station in Laurel. *[William A. Wulf Collection]*

Laurel train station in decay. *[William A. Wulf Collection]*

Glenwood

Glenwood was a small settlement about five miles north of Scotts Valley, on Glenwood Highway. California Historical Landmark Number 449 is located in front of the former train depot site, across from Martin's store and the Post Office. It reads:

Historic town founded by Charles C. Martin, who came around the Horn 1847, and his wife, Hannah Carver Martin, who crossed the Isthmus. First homesteaded the area in 1851 and operated toll gate and station for stage coaches crossing Mountains. Later Martin developed lumber mill, winery, store and Glenwood Resort Hotel [State Registered Landmark No. 449. Tablet planned by California Centennial Commission. Base provided by Santa Cruz Parlor No. 26, Native Daughters, Santa Cruz Parlor No. 90 Native Sons, and descendants Martin Family. Dedicated June 22, 1950]

After arriving in California, Martin first worked as a teamster in the lumbering area near Lexington and then in 1851 homesteaded land adjoining that of Mountain Charlie McKiernan. Together with McKiernan, he first operated a toll gate and stage station at Station Ranch. The Martins soon found themselves with six children and a need for more income.

They built a store they called Martinsville in 1870 (or possibly 1873), renaming it later as Glenwood. Apparently the name itself comes from the fact that the community was located in a wooded glen.

44

Over a period of years, Martin developed businesses alongside the Mountain Charlie Road between Scotts Valley and the Summit area and Glenwood became a popular tourist spot. Besides the retail establishments, there was a popular health spa known as the Glenwood Magnetic Springs. The water was supposed to have medicinal properties because it passed over magnetic iron ore. After Martin developed the Glenwood Hotel, he sold it to a Catholic Sisterhood.

Mrs. Ed C. Koch, great granddaughter of C.C. Martin, founder of Glenwood, at the memorial tablet dedicated in 1950. Mrs. Koch has lived her entire life in Glenwood, except for a few years away at school. *[UCSC Special Collections]*

From a postcard advertising the Glenwood Hotel. *[UCSC Special Collections]*

Famous Magnetic Springs at Glenwood. The springs were believed to have medicinal powers. *[UCSC Special Collections]*

In 1880 the Southern Pacific Coast Railroad (narrow gauge) reached Glenwood. Martin wisely donated a Glenwood Train Station and soon the train had a regular stop there, after it had emerged from a series of tunnels when travelling southbound. The train stop greatly increased the public's access to the area.

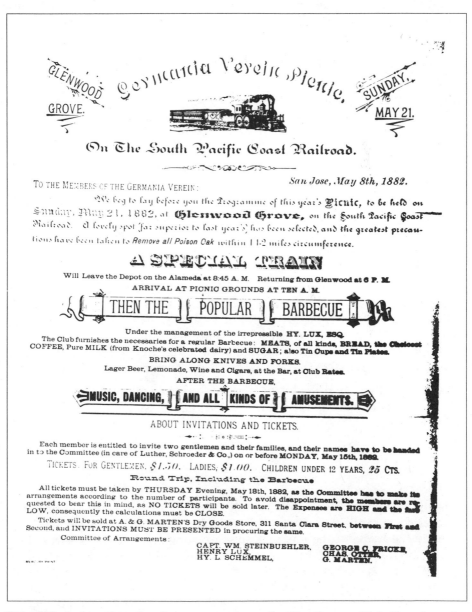

This 1882 picnic notice advertises an event for an East Bay German Club. Travel was by a special train from the Alameda to Glenwood. "A lovely spot has been selected and the greatest precautions have been taken to remove all poison oak." Excursions like this were popular at the time. [*William A. Wulf Collection*]

When the California State Engineers began to talk about building an auto road through the Santa Cruz Mountains, they found that Charlies Martin had anticipated them. He had a survey run and paid for it himself. Obviously the best route was through Glenwood and that's where the road went when it was built in 1916. When it was paved with concrete in 1919, Charlie was asked to put his name and footprints in the new surface......in 1934 the highway was realigned and Highway 17 was built to take a course completely away from Glenwood. The Martins' store and gas station closed the same year. [Margaret Koch, p. 142]

Unfortunately in recent years, during a re-paving job, Martin's footprints were covered over. Margaret Koch, the Scotts Valley historian and writer, is the great-granddaughter of Charles Martin.

With both the railroad closing and the new Highway 17, Glenwood's days were numbered. The town's last post office was finally discontinued in 1954.

Saint Clare's Retreat House

Southbound travelers may see the Saint Clare's sign just before entering Scotts Valley. Saint Clare's is a religious retreat house run by the Franciscan Missionary Sisters of Our Lady of Sorrows. Twelve sisters maintain the property and offer weekend spiritual retreats for women in a beautiful secluded mountain setting.

The retreat center was named to honor Saint Clare, the first woman Franciscan in Assisi. The order was started in the 1940s with the mission of bringing its religious message to China's large population. They were expelled from China during World War II and returned to California, purchasing the property in 1950. Originally the site was the Mountain View Ranch resort, one of four similar resorts in the area. All of the original buildings have been replaced.

St. Clare's is located at the intersection of Rodeo Gulch and Mountain View Roads, off of Branciforte. The address is 2381 Laurel Glenn Road, Soquel, (408) 423-8093.

Truck Farm

What is often popularly referred to as the "truck farm" is on the west side of the highway, just south of the Scotts Valley city limit sign at 7260 Highway 17. It is actually the home of the Roy C. Davis Trucking company, not surprisingly owned by Roy Davis. This is a private business which does not perform car repairs. Mr. Davis has owned the property for over 20 years and has the distinction of owning the only billboard sign on Highway 17. Although billboards are illegal along highways using federal funds, Mr. Davis had obtained legal authorization from the city before the federal law came into effect.

Santa's Village

Built in 1957, this amusement park was very popular with tourists. Built around a North Pole theme, the park featured rides, picnic areas and of course Santa himself. Originally the site was a dairy farm.

In 1990 Borland International purchased the land as the site of its world headquarters. Borland is a $500 million dollar a year company that produces software for the personal computer market.

Scotts Valley

The City of Scotts Valley is relatively newly - incorporated in 1966 and encompasses five square miles of land. Before incorporation as a city, the general valley area was widely known as Scott's valley, named after Hiram David Scott who purchased land there in 1850 for $25,000.

Scott, born in Pittston, Maine, on January 28, 1822, became a sailor and in 1846 while serving as second mate of the sailing vessel J.C. Whiting, jumped ship in Monterey; he settled in Santa Cruz, became a shipbuilder, but left for the "mines" in 1848....he returned to Santa Cruz county in 1852 and settled on Rancho San Agustin which he had purchased from Joseph Majors two years earlier. [Clark, p.331-2]

In 1865 a school district was created and soon after dairy farms began in the area. In 1950 Bethany Bible College, a four year college, moved from San Francisco purchasing property just west of Highway 17. Today the city boundaries are just north of the truck farm (above the former Santa's Village) and on the south

Santa's Village. "Welcome House is the entrance to a child's paradise in a fairyland forest setting" boasted the advertising brochure. *[UCSC Special Collections]*

Hiram Scott was neighbor to Mountain Charlie and helped build the toll road in 1858. Scotts Valley is named after him. *[William A. Wulf Collection]*

The Scotts Valley School House in 1916. The teacher is in the back row on the left. Ruby Strong, who provided the photograph, is in the front row on the right. This was the entire school population in Scotts Valley. *[Courtesy of Ruby V. Strong]*

50

The Frapwell property in Scotts Valley looking east across the valley. Today Highway 17 runs through the middle of the picture from left to right. The new Borland office complex will be to the left of the buildings. This picture was taken in 1926 by Elvis Frapwell. *[Courtesy of Ruby V. Strong]*

Camp Evers store marked the corner of Scotts Valley Drive and Mount Hermon Road. This picture was taken in the 1930s. *[William A. Wulf Collection]*

where it meets the Santa Cruz city limit before the southern end of Highway 17. Population today is 8,600 and the dairy farms have given way to housing and light industry.

The Tree Circus

One of the most remembered spots in Scotts Valley was Axel N. Erlandson's (1884-1964) Tree Circus business. Located along Scotts Valley Drive (the old road to Santa Cruz), Axel had pruned 55 trees into exotic shapes.

Axel Erlandson with one of his specially pruned trees. *[UCSC Special Collections]*

Over 55 trees graced the Tree Farm in Scotts Valley, each with a unique shape. *[UCSC Special Collections]*

Erlandson was a farmer of Swedish descent who was farming near Turlock in the 1920s. One day he noticed how two tree branches had grafted themselves, creating a new form. He began experimenting with creating trees of strange and unique shapes.

In 1945 he bought land in Scotts Valley and two years later retired there, moving some of this trees. Soon he had 55 sycamores, willows, box elders, ash, poplars and maples that had been pruned and graphed into "spirals, rings, loops, crooks, hearts, igloos, ladders and chains." Axel put up a sign "See the World's

Strangest Trees Here" by the side of the road and soon turned it into a tourist business (admission was 30¢). He personally conducted tours of his unique creation. Later he built a castle-like building complete with turrets to attract attention. Ripley's Believe it or Not newspaper column featured the Tree Circus in one issue, as did Life magazine.

In 1963 Erlandson sold the Tree Circus for $22,000 and he died a year later. It passed through four different owners and from 1967-73 the yard housed huge plastic dinosaurs as part of an attraction called "Lost World."

In 1987 Michael Bonfante, owner of Nob Hill Foods, bought the trees and transported 28 of them to a nursery yard in Gilroy. He is building a "enchanted forest" theme park in Gilroy and will use the trees there.

Santa Cruz architectural designer Mark Primack gets the credit for raising public awareness about the Tree Circus and preserving its unique story.

Santa Cruz

Santa Cruz is both the name of a county and a city (which happens to be the largest city in the County), located 37 miles south of San Jose. Portola "discovered" what is presently Santa Cruz the same year that Spain claimed California (1769). The name means "holy cross" in Spanish. Mission La Exaltacion de la Santa Cruz was established in 1791 by Father Francisco De Lasuen as the 12th mission in "Alta California." The first stone for the five foot thick rock and adobe walls was actually laid in 1793. Eventually other buildings were added around a square that remains today, along with a rebuilt replica of the original church. The original building was weakened by earthquakes and destroyed in 1857. The

Painting of the original Santa Cruz Mission established in 1791. [*Covello and Covello Photography*]

Early view of Santa Cruz from High Street. The ocean is visible at the top of the picture.
[Covello and Covello Photography]

current replica of the Holy Cross Church met the same fate as a result of the 1989 earthquake and is being re-built.

Historical writer Major Rolin Watkins records in 1925 that area travel in the 1850s was entirely on horseback (a caballo). Husbands and wives rode the same horse with the husband behind, holding the reins. They generally followed trails, but there were no bridges. Streams swelled in winter and often became un-fordable. No one was in a hurry, time was not an essence of any contracts. All large ranches were debt free and stores gave easy credit to the wealthy landowners - but eventually those bills came due and many were unable to pay. Eventually most of the ranches passed into the hands of local merchants. The towns were run quite democratically, the concept of a "prominent" citizen had not yet made a presence. The population was given as 640 in 1850 - but at that time non-anglo citizens were not counted for census purposes.

What we know today as the City of Santa Cruz came into being around 1830 when a village was formed around an early Catholic Mission located near Santa Cruz Creek. In 1866 the town was officially recognized by the state of California with boundaries very close to what still exist today.

In 1915, when the first roads for automobiles were being started, Santa Cruz was quite different from today. 5-10 ships were commonly anchored at one of the three piers, delivering supplies to the area. The train station was very busy with

The northern Santa Cruz City limit sign on Plymouth Street, at the corner of Highway 17 in the 1930s. *[UCSC Special Collections]*

several trains a day arriving. In 1910, one either took advantage of the excellent train service or one of the passenger ships which serviced the community three times a week.

In 1912 the Santa Cruz Board of Supervisors authorized "the hiring of a man who could ride a motorcycle to protect the county roads against law breakers who drove their automobiles over the 20-mile speed limit." [Santa Cruz Sentinel, September 3, 1935, p. 4]

Santa Cruz remains an agriculturally based county, with tourism playing an important part in the economy, along with the University of California campus established in 1965. A major earthquake in 1989 caused extensive damage to the downtown Santa Cruz business area buildings, but the town is rebounding.

Today the County contains 440 square miles of mountains and plains that border on the Monterey Bay. Estimated population in 1988 was 225,500. Agriculture still plays an important role with apples being the primary cash crop. Tourism became important in 1900 as people became more mobile and desired to visit the beautiful beaches of the Monterey Bay.

Margaret Koch's book *Santa Cruz County, Parade of the Past* is an excellent source of more history about Santa Cruz.

Soquel

The name Soquel, according to tradition, is a Spanish rendering of an Indian name, but the meaning has been lost. The site of today's village was a ranch located near the Soquel Creek. Leon Rowland states that Soquel most likely

Soquel in the 1880s. The church in the foreground still stands today. The Soquel bridge is in the middle of the picture and the old Santa Cruz Highway can be seen horizontally in the middle right of the picture. *[Covello and Covello Photography]*

came from an Indian term meaning "place of the willows." The town itself was founded in 1852.

Soquel was the hub of the county until the mid-1900s. There was considerable lumber activity above Soquel, and a major mill was located in the town. Also one of the major trails to Los Gatos started in Soquel making it a focal point for business. By 1860 it was the third largest community in Santa Cruz County.

Highways 1 and 17 changed that and today Soquel is regarded as a quiet residential community. It was officially incorporated on January 11, 1949.

The popular Senate Saloon was at the corner of Soquel Drive and the old Santa Cruz Highway in Soquel. Sam Alkire was the owner (shown in front of the saloon about 1900). *[UCSC Special Collections]*

CHAPTER 3

Early Transportation Routes

Early Government Support of Roads

The idea of government roads first came from Thomas Jefferson. Jefferson had a fear of strong centralized government and made decentralization of the communications and transportation industries one of his primary missions during his term as President (1801-09).

During the War of 1812, Andrew Jackson found it very difficult to move troops on the poorly maintained roads so for the first time the Federal Government started discussing a national road system. The idea of *building* roads was still so strange in a nation of trails that Secretary of the Treasury Albert Gallatin described them in his *Report on Roads and Canals* in 1808 as "artificial" roads. Under the 1921 act establishing the federal primary aid system, justified to insure good roads in the case of war, up to 7% of a state's highways could be designated as "federal" and thus receive additional funding and protection.

Around 1900 California began serious discussions on building a network of roads inter-connecting counties to facilitate trade. But the arrival of the automobile a short time later changed California forever and made roads a primary concern of governments.

The First Santa Cruz Mountain Trails

We begin with the first trails over the mountains. Travel in the Santa Cruz Mountains before the 1850s meant walking or trying to ride a horse through the scrub brush. One of the first families to settle in the Summit area, the Schulthesis family, wrote that it took them three travel days to go from Los Gatos to the summit, using oxen to break a trail through heavy brush.

Father Fermin Francisco De Lasuen is credited with being the first outsider to transverse the Santa Cruz mountain range, while performing his duties establishing the mission in Santa Cruz.

Father Lasuen was born in the Basque province of Alava, Spain and founded nine of the twenty-one California Missions. In a 1791 report written at the San Carlos Mission, he records:

I proceeded to Santa Clara in order to examine anew in person the site of Santa Cruz. I crossed the Sierra by a long and rough way, and I found in the site the same excellent fitness that was reported to me. I found, besides, a stream of water very near, copious and important. On the day of San Agustin, August 28, I said mass, and a cross was raised in the spot where the establishment is to be. Many gentiles came, large and small, of both sexes, and showed that they would gladly enlist under that sacred standard, thank God! I returned to Santa Clara by another way, rougher but it was also shorter and more direct. I made arrangements for repairing it by means of the Indians of the mission I just mentioned; and an excellent job has been done because for that job, as for all others, the Commandant of the Presidio of San Francisco, Don Hermenegildo Sal, furnished all that he was asked, with all speed and promptness. [Kenneally, p 235.]

Historian William Wulf has studied this period of our history extensively and believes that the exact date of Father Lasuen's journey was August 28, 1791. This latter new road, really a trail, linked the two missions and the first regular trade began between the two areas. This also marks the beginning of the formal road system that eventually led to Highway 17.

In 1795 the newly appointed Diego de Borica, Governor of California, ordered improvement of the road. Payne records that:

On December 31, 1799, Governor Borica ordered the Branciforte settlers not to use the mountain road for pleasure trips to San Jose without specific advance approval. The Governor wanted the settlers to work their land rather than loiter about in San Jose. The road, in the Governor's view, would be used only for the limited purpose of bringing supplies... [Payne, *Howling*, p. 11]

In 1846 Captain John Charles Fremont explored the area from Los Gatos to Santa Cruz mentioning the road that existed. Early settler Lyman John Burrell in his memoirs records:

No man had ever been known to drive over the summit with a wagon. It was considered not difficult, but a rather dangerous undertaking. In those days, a man could not safely travel very far alone, unless he was well armed, because bears were not unfrequently seen on the trails, and they had not always the politeness to turnout for a man; but on the

Early Roads Map

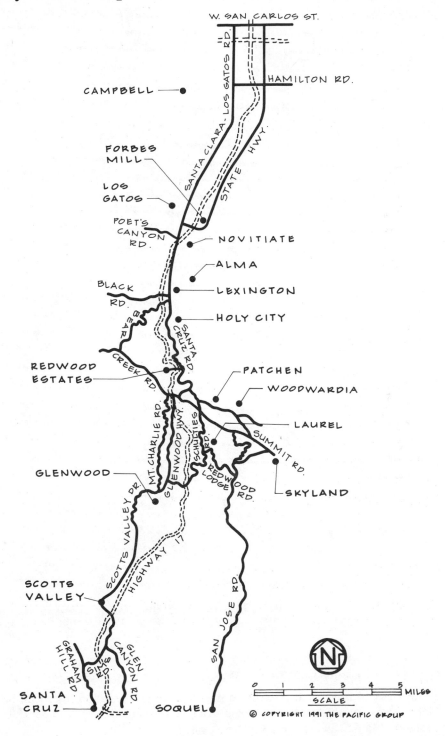

W. SAN CARLOS ST.

HAMILTON RD.

CAMPBELL

SANTA CLARA - LOS GATOS RD.

STATE HWY.

FORBES MILL

LOS GATOS

POET'S CANYON RD.

NOVITIATE

ALMA

BLACK RD.

LEXINGTON

HOLY CITY

BEAR CREEK RD.

SANTA CRUZ RD.

REDWOOD ESTATES

PATCHEN

WOODWARDIA

LAUREL

SCHULTIES RD.

SUMMIT RD.

GLENWOOD

MT. CHARLIE RD.

GLENWOOD HWY.

REDWOOD LODGE RD.

SKYLAND

SCOTTS VALLEY DR.

HIGHWAY 17

SAN JOSE RD.

SCOTTS VALLEY

GRAHAM HILL RD.

SIMS RD.

GLEN CANYON RD.

SANTA CRUZ

SOQUEL

N

0 1 2 3 4 5 MILES
SCALE

© COPYRIGHT 1991 THE PACIFIC GROUP

Page of the handwritten report (in Spanish) sent by Father De Lasuen that records the first road over the Santa Cruz Mountains. [*Courtesy of Mission Santa Barbara*]

contrary, they would sometimes dispute his passage. The poor conditions remained even after the road to Santa Cruz was improved in the late 1850s. [Payne, *Harvest*, p. 105]

From Los Gatos north, there were trails towards San Jose on either side of Los Gatos Creek; on the west side the route of today's Winchester Blvd., and on the east side Main St. which turns into Bascom Ave.

Early Toll Roads

There were three basic parts to the early toll roads. From Los Gatos travellers went south to the summit area. They then had a choice of going slightly east and taking the road to Soquel or going west to Mount Charlie Road, which continued through Scotts Valley into Santa Cruz. These basic transportation patterns from the 1850s continued until Highway 17 was built in the 1930s.

The earliest trails were informal but in the early 1800s, the date is unclear, "Buffalo" Jones turned the trail through his property in the Lexington area into a toll road and named it "Farnham's" in honor of Mrs. Eliza Woodson Farnham who was the first woman to cross the Santa Cruz Mountains in a buggy accompanied by a ranch handyman. The trail was so steep in places that the buggy had to be disassembled and carried by hand.

Mrs. Farnham, a writer, social worker, feminist and activist was a fascinating woman and representative of the pioneer women who inhabited the area along Highway 17. Born in 1815 and originally from New Jersey, she helped reform the prison system for women. She was widowed in 1848, her first child died and this tragedy was soon followed by a second divorce. (You can read more about her life in Joan Barriga's excellent paper on *The Women of the Santa Cruz Mountains.*)

The toll road was the only path between much needed lumber in the mountains and the growing demand in the Santa Clara Valley. Jones understood the value of his monopoly and for years he collected tolls from every passing wagon, stage and individual rider.

In 1846 Buffalo Jones is quoted:

Steep, rough and in the summer time a wondrous place for dust, the trail in winter time was a dangerous place for pack trains and men alike. The bull teams that followed and the stage coach that came later with their iron shod wheels did little towards improving the route, simply transforming it into a pair of parallel runs that provided a hair raising ride for venturesome travellers [Payne, *Howling*, p. 12]

These roads were 6-7 feet wide with a untreated dirt bed. Winter rains turned the roads to mud. Dust from the traffic was a major complaint by area residents. Again Payne records Summit resident Lyman Burrell travelling over the road in 1853:

The ascent of the mountains was not as easy in those days as it is now. We had then no graded turnpike. The road we were to travel had been made for the purpose of getting down logs. It was very tough and steep, and sometimes very sideling. In some places we found it difficult to keep the cattle from sliding off the lower side. We first went over Jones's hill, a distance of about 4 miles, on the East side of the Creek; then we crossed over and went to the top of another hill on the north side of Moody's gulch....On climbing these hills we had to double our team, and carry up one load at a time. It was so rough and steep that we had to partly unload our wagons and take up only a part of a load at a time, thus making several trips. [Payne, *Howling*, p.12]

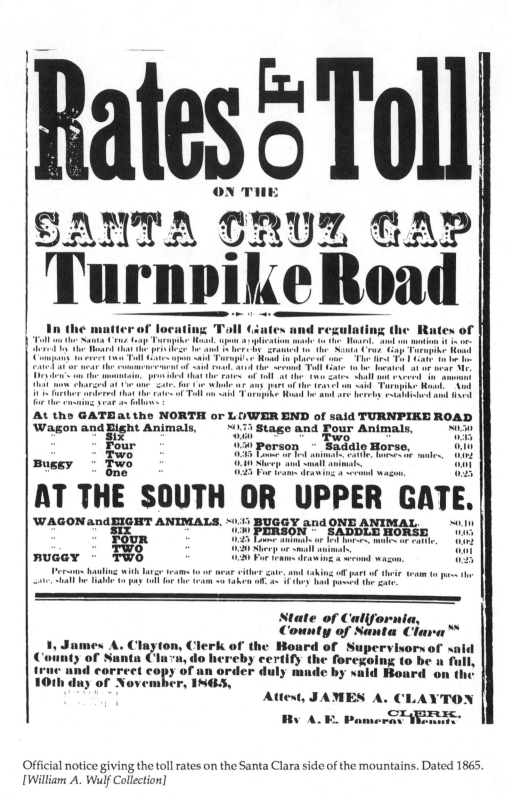

Rates ᴏꜰ Toll

ON THE

SANTA CRUZ GAP
Turnpike Road

In the matter of locating Toll Gates and regulating the Rates of
Toll on the Santa Cruz Gap Turnpike Road, upon application made to the Board, and on motion it is ordered by the Board that the privilege be and is hereby granted to the Santa Cruz Gap Turnpike Road Company to erect two Toll Gates upon said Turnpike Road in place of one The first Toll Gate to be located at or near the commencement of said road, and the second Toll Gate to be located at or near Mr. Dryden's on the mountain, provided that the rates of toll at the two gates shall not exceed in amount that now charged at the one gate, for the whole or any part of the travel on said Turnpike Road. And it is further ordered that the rates of Toll on said Turnpike Road be and are hereby established and fixed for the ensuing year as follows :

At the GATE at the NORTH or LOWER END of said TURNPIKE ROAD

Wagon and	Eight	Animals,	$0.75	Stage and Four Animals,	$0.50
"	Six	"	0.60	" " Two	0.35
"	Four	"	0.50	Person " Saddle Horse,	0.10
"	Two	"	0.35	Loose or led animals, cattle, horses or mules,	0.02
Buggy	Two	"	0.40	Sheep and small animals,	0.01
"	One	"	0.25	For teams drawing a second wagon.	0.25

AT THE SOUTH OR UPPER GATE.

WAGON and	EIGHT	ANIMALS,	$0.35	BUGGY and ONE ANIMAL.	$0.10
"	SIX	"	0.30	PERSON " SADDLE HORSE	0.05
"	FOUR	"	0.25	Loose animals or led horses, mules or cattle,	0.02
"	TWO	"	0.20	Sheep or small animals,	0.01
BUGGY	" TWO	"	0.20	For teams drawing a second wagon.	0.25

Persons hauling with large teams to or near either gate, and taking off part of their team to pass the gate, shall be liable to pay toll for the team so taken off, as if they had passed the gate.

State of California,
County of Santa Clara ss

I, **James A. Clayton, Clerk of the Board of Supervisors of said County of Santa Clara, do hereby certify the foregoing to be a full, true and correct copy of an order duly made by said Board on the 10th day of November, 1865,**

Attest, **JAMES A. CLAYTON**
CLERK,
By A. E. Pomeroy Deputy

Official notice giving the toll rates on the Santa Clara side of the mountains. Dated 1865.
[William A. Wulf Collection]

Frederick A. Hihn, Father of Capitola and investor in both the Soquel and Mountain Charlie toll roads. *[Courtesy of Rick Hamman]*

Improved Roads

Local residents were upset at the toll road, claiming Jones charged exorbitant tolls and had made the road so steep that lumber wagons could only haul half a load (thus doubling his income). Merchants began to talk about making a new improved road over the hill, leading to the true early ancestor to Highway 17.

Under pressure from teamsters and lumbermen, the Santa Clara Board of Supervisors appointed Charles White, P.J. Davis and A.S. Finlay to study a possible route. Their 1852 report concludes:

Your Commissioners believe that a road to Jones' Mill would be very beneficial to a portion of the people of Sta. Clara County for the purpose of hauling timber there, but as there is a considerable portion of the county which would derive no benefit from such

This picture was taken above Alma in the 1880s near Black Road. Behind the four horses is the "Fresno" road scraper used at that time to clear and level roads. *[Courtesy William A. Wulf (from the William Van Lone collection)]*

a road, what so ever, we are of opinion, that it would be unpublic, for the county to open said road any further than Major Chases, to which place it will be beneficial to those living in that section and for hauling timber, and as a road by horseback to Santa Cruz, and the cost of so opening it will be comparatively nothing, as there will be no person injured by it, in our opinion, but the public will be greatly benefitted as there is much travelling that way to Santa Cruz by horseback. [Wulf, *History of Santa Cruz Mountains*, p. 41]

The State of California created the Plank and Turnpike Roads Act in 1853 that allowed private groups to organize joint stock companies to build private toll roads, stipulating that after 20 years the roads would become public property.

In 1854 the Santa Clara County Board of Supervisors authorized a road but there were no funds available to build one, so in 1858 a committee of representatives from each county was formed and a trio of road viewers, headed by Sheriff John Murphy, was sent out to "view out" (survey) the entire route of a possible road. The group agreed that a "Los Gatos-Santa Cruz Turnpike" toll road was the best solution with the trail crossing the County line at the Santa Cruz Gap, now known as Patchen Pass. Construction for each side would be the responsibility of the respective county.

The Santa Clara group, organized in 1857 as the Santa Cruz Gap Turnpike Joint Stock Company, bought the Zachariah "Buffalo" Jones toll road and began upgrading the route at a cost estimated at $7,000. John Roork, Rufus Herrick, James Howe (Secretary), D.D. Briggs, E. Wilcox, John S. Favor, Stephen I. Easley (President) and Alexander Hogan made up the Committee. Historian William Wulf states that Zachariah Jones was unhappy with the new arrangement and his loss of tolls. He threatened to sue the County and void the agreement, but eventually sold his property. The road was finished in October 1858.

On the other side of the mountain, in 1858 the Santa Cruz Turnpike Company was formed by several Santa Cruz County residents to create a toll roll from Soquel to the summit. Local Historian Leon Rowland records the first meeting of the group was held on January 30 headed by Judge Henry Rice, Eliah Anthony (merchant, Postmaster in Soquel), Samuel A. Bartlett (banker, furniture dealer), Nathaniel Holcomb (lumberman), F.A. Hihn (father of Capitola) and James Hames (mill owner). Frederick Augustus Hihn was a wealthy landowner with a logging business and mill, apple sheds and a bank. Besides founding Capitola, he is also known as the person who at one time paid one tenth of all the Santa Cruz County taxes! He was born in Germany in 1829 and came to Santa Cruz in the 1850s.

The Santa Cruz County portion of the road cost $6,000. In 1860 the group re-organized as the Santa Clara Turnpike Company.

The new 8' wide dirt road ran south up the west side of Los Gatos Canyon, then to the old Jones Hill stage road through Lexington up to the summit where it met the Santa Cruz road that basically followed the route of today's Soquel-San Jose Road, west of the Soquel River, ending in Soquel. Officials called the road the "Old Santa Cruz branch of the El Camino Real."

Road building was largely by hand, with some help from metal scrapers hauled by mules or horses. The scrapers performed the same function as today's road graders which have a blade that digs slightly into the surface, bringing up loose dirt which falls into the holes - hopefully resulting in a more even surface.

The new road was first used on May 5, 1856 by Joseph Johnson and Peter Davidson who successfully drove a wagon from San Jose to Soquel.

...a moderate grade allowed horses to trot the seven miles to the summit. The road was cut into the side of steep hills and was so narrow and crooked that turnouts were provided to allow wagons to pass. Although these were an improvement, many accidents...occurred with unmanageable horses and careless drivers, resulting in the loss of both horses and wagons. [Payne, *Howling*, p. 15]

Portions of Mountain Charlie Road still exist today near the summit. *[Richard A. Beal]*

67

Scotts Valley Toll Road House, home of Hiram Scott. The finished building lumber was brought around the horn in 1853. At the time Highway 17 was built in the 1930s it was moved to the area of the Scotts Valley City Hall. *[San Jose Historical Museum]*

Pacific Ocean House (hotel) in Santa Cruz with the stage out front. This photograph is from 1/2 of a 1866 stereoptican held by the California Pioneer Society Museum. *[William Wulf Collection]*

Parkhurst drove stage coaches like this one in Felton. The road dirt left everyone, passenger and drivers like, dusty at the end of a trip. *[Covello and Covello Photography]*

While a great improvement, it was hardly like today's road. The summit to Soquel section crossed Soquel Creek 25 times without any bridges! Santa Cruz County Board of Supervisor minutes record the tolls as 50¢ for a wagon and two horses or oxen, 37 and one-half cents for a horse and buggy, 25¢ for a horse and rider, 5¢ for each animal. Some Santa Cruz locals called it the Paper Mill Road, referring to a saw mill one-half mile north of Soquel owned by Edward and Frank O'Neill.

At about the same time, a rival effort was started by Mountain Charlie McKiernan. McKiernan had many trails and roads through his property, the most used one being a cut-off down the old Indian trail near his home at the summit, through the Moody Gulch territory to Los Gatos Creek. Later he scratched a crude road through his property from Redwood Estates to the summit, then through to Glenwood, using a "V" scraper. He and his neighbor, Hiram Scott, established toll gates at his cabin home one-half mile west of the summit, and at the other end one-half mile from the Glenwood Hotel. This road was officially the Santa Cruz Gap Turnpike but was known popularly as Mountain Charlie Road. A trip from Scotts Valley to Santa Clara with a horse and buggy took 5 hours in 1858.

The Soquel group finished their improved road first but both roads were widely used.

The Santa Cruz based *Pacific Sentinel* claimed, in an article about the 1862 official opening:

A Glorious Triumph of Industrial Improvement. We can announce to the world, or more especially to the citizens of Santa Cruz and Santa Clara counties, the completion of the Santa Cruz Turnpike, as far as contracted for, by the industry of [Mountain Charlie] McKiernan and [Hiram] Scott, who have nobly finished, by strong effort, a road to the dividing summit, and made it passable and safe for carriages and teams, with ordinary prudence on mountain roads...And we have satisfaction of announcing, that stage coaches run triweekly, over the road. [Wulf, History of Santa Cruz Mountains, p. 52]

Stage Coach Lines

The first stage service between San Jose and Santa Cruz began in 1854. The only stage route went through San Jose, San Juan Bautista, Watsonville and then north to Santa Cruz. Travellers from San Francisco took two days to reach Santa Cruz. One way fares from San Jose to Santa Cruz were $4.00.

By October 1858, the new toll roads over the Santa Cruz Mountains shortened stage travel to a one day trip from San Jose. The first stages used Mountain Charlie Road, but soon there were several competing stage lines using the two different routes (Soquel or Mountain Charlie) with service daily. Fares were dropped to $2.50 one way.

One of the most famous stagecoach drivers was Charley "Darkey" Parkhurst, known as one of the toughest and most reliable drivers on the line. He lived for

Although based in San Jose, the Elson stage line ran through the Santa Cruz Mountains. *[San Jose Historical Museum]*

20 years around the Freedom Blvd. and Day Valley Road area in Aptos. Wearing a black patch over one eye, he was referred to as "one eyed Charlie." However Parkhurst is best known for having living her life as a man, until her death when her true sex was revealed for the first time to startled friends.

Many people have confused Mountain Charlie McKiernan with Charley Parkhurst. Although both were colorful figures living at about the same time, they were quite distinct individuals.

Born Charlotte Parkhurst in 1812 and orphaned at an early age, she apparently escaped the orphanage dressed as a boy — a custom she continued throughout the rest of her life. She found work as a stable boy and eventually came west in 1851 during the California gold rush. She worked as a stage coach driver and was well known in the area.

She smoked cigars, chewed tobacco, drank moderately, played cards and shook dice for cigars or drinks. [Barriga, *Women*, p. 24]

In 1868 Ulysses Grant was running for President and Parkhurst went to Soquel where she registered to vote - making her the first woman to vote in California! In 1955 the Pajaro Valley Historical Association placed a gravestone in the Watsonville Cemetery on the West side of Freedom Road:

Noted whip of the gold rush days drove stage over Mt. Madonna in early days of Valley. Last run San Juan to Santa Cruz. Death in cabin near the 7 mile house. Revealed "one eyed Charlie" a woman. First woman to vote in the US November 3, 1868.

Another old stage road to San Jose left Santa Cruz by fording the San Lorenzo River, where the River Street bridge is now. Then up Graham Grade, passing over what is now the Pasatiempo Golf Course and onto the first stop, the ranch of Abraham Hendricks in Scotts Valley where two horses were added for the long pull ahead. Then up to Mt. Charlie's station and eventually down into Santa Clara Valley. Apparently Mountain Charlie McKiernan owned the line until 1874 when it was sold to George Colgrove. They used a popular yellow Concord coach with leather springs. In the late 1800s a "mud wagon," all terrain vehicle was put in use as a replacement for the much heavier Concord coaches.

Popularity Demands Better Roads

The roads were still quite rough and in 1861-2 heavy rains damaged much of the route. By 1890 the summit area began to decline in importance. The forests had been clear cut, the hilly area made agriculture difficult, and the new popularity of the automobile meant that people were more willing to travel further distances. The summit was no longer attractive as a wilderness spot.

In 1896 Phil Francis wrote a tourist guide book to Santa Cruz and mentioned the roads: "All [Santa Cruz] county roads are built scientifically and sprinkled daily. [Santa Cruz has] the best roads in the state as certified by the State Road Commission" [Francis, p 4.]. Describing the road from Scotts Valley, he writes:

At the head of the valley [Scotts] the 'Old San Jose Road', that first thoroughfare across the mountains into the Santa Clara Valley, begins to ascend that mountain; the hills and the big trees close in and, finally, the way climbs to the crest of Vine Hill. A branch road leads to Glenwood Magnetic Springs, another to beautiful Summer Home Farm, the Bernheims vineyard and orchard place; still others connect, by picturesque routes, with several of the stations on the narrow gauge railroad with B.C. Dakin's Lauren Glen Fruit Farm and with Mrs. T. P. Robb's slightly Sea View Villas. [Francis, p. 121]

The Soquel-San Jose Road (old Santa Cruz Highway):

Nothing could be more romantic and picturesque than the drive of sixteen miles which leads from the minimal springs near the headwaters of the Soquel and sticks close as a brother to that stream, which tumbles down, broadening as it goes and cutting its way among the big redwoods, and the trickle of the same name, and from there flows demurely and calmly out into the bay at Capitola. The road is a pretty good one, as mountain roads go; in fact when you reflect that it crosses the creek 25 times in its brief career, and most frequently without the aid of a bridge, you are disposed to regard it as an astonishingly good road under the circumstances. [Francis, p. 122]

But it still took two strong horses a long day to complete the journey between Los Gatos and Soquel, and overnight stays were common. Margaret Petsch wrote in the Santa Cruz Sentinel:

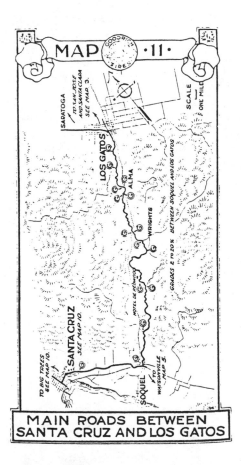

SANTA CRUZ AND LOS GATOS.

Between Santa Cruz and Los Gatos is a wild mountainous country. GOODRICH ROAD MARKERS will be found all the way.

The road here climbs to the very crest of the coast range, affording many good views of mountain scenery. Near the summit of the divide are a number of hotels and resorts, where one may spend a few days. With a good strong car, many pleasant side roads may well be explored.

This mountain road is used as the main route between Santa Cruz and San Jose or points in the vicinity of San Francisco. Tourists who wish to avoid the mountains must go via Gilroy or Half Moon Bay.

HOTELS AND GARAGES: Santa Cruz, Soquel and Los Gatos. Hotels are also found at points in the mountains.

ROAD CONDITIONS: Usually fair, with some dust in the summer. After a heavy rain or during the winter, clay may make the grades dangerous. At these times, the detour via Gilroy should be taken. There are grades up to 23%, and a number of sharp turns. Caution should be used, as cars are sometimes driven at high speed through here.

ROAD MILEAGE: Santa Cruz to Soquel, 3.1 miles. Soquel to Los Gatos, 25 miles. Los Gatos to San Jose, 10 miles. San Jose to Santa Cruz, 38.1 miles.

GOODRICH TIRES: At Santa Cruz and Los Gatos.

NOTE: For touring north of San Francisco, see later GOODRICH ROUTE BOOKS of the Pacific Coast.

Goodrich Tire Company produced automotive guide books in the early part of the century, like this one describing the road between Los Gatos and Soquel. They also installed advertising signs, known as Goodrich Road Markers, to help people find their way. [*Bruce Kennedy Collection*]

The width of the road allowed for one-way traffic and those who have made the trip by wagon can remember the most unpleasant experience of having to back their horses down hill upon meeting another conveyance unexpectedly. [*Santa Cruz Sentinel*, September 5, 1935, p. 4]

Roads For Automobiles

In 1897 the California Legislature dissolved the Bureau of Highways and created a Department of Highways. Marshden Manson, J.R. Price and W.L. Ashe shared management for the first three years and toured Europe to study new road building techniques. They get credit for envisioning the first paved road system in the state. In 1902 the California Constitution was amended giving the Legislature the power to set up a system of state highways.

The existing "roads" in the Santa Cruz Mountains were gradually being improved and automobile traffic increased, although automobiles were still rare.

In the early 1900s Mrs. Arthur Sears initiated a petition to ban automobile traffic from the mountain roads. She claimed that automobiles were a hazard and frightened horses, making driving a buggy unsafe. [Payne, Howling, p. 16]

However, government officials realized the automobile was here to stay and in 1905 began a program to improve all major roads to permit automobile traffic. The 1906 earthquake, however, did extensive damage to the road and delayed its upgrade. Travel had become so popular on the old road that the gravelled surface proved inadequate to carry the traffic. The road was one of the first designated by State highway engineers to be covered by Portland cement.

In 1909 the California State Highway Act was passed, authorizing the Department of Engineering to map a State Highway System and in 1913 registration of all motor vehicles was required. Santa Cruz desired to be linked to the outside world by automobile, and on September 11, 1911 the Santa Cruz Chamber of Commerce passed a resolution that :

A County Highway Commission select the best shortest and most available route for the Santa Clara [county] line over the Santa Cruz Mountains via Los Gatos Canyon, to Santa Cruz and thence to Watsonville. [Olin, p. 63]

A 1911 $18 million State bond issue required every County seat to be linked to several designated main arteries. These roads were called "county seat laterals." By 1912 California had the first highway in America specifically constructed for the automobile. The road between Los Gatos and Santa Cruz was one of the first marked for such upgrading. Later bond issues in 1915 for $15 million, 1917 for $18 million, and 1919 for $40 million financed the state's initial automobile road system. In 1921 the current method of road construction financing, the gasoline tax, was introduced and started at 2¢ per gallon. Today it is over 31¢ per gallon!

Improving roads was a very popular item with taxpayers of the time. The 1919 bond vote in Santa Cruz County, for example, was 2,872 for and only 170 against. But finding purchasers of the bonds was difficult and took considerable time. In fact, Santa Cruz County had to purchase $22,000 of the state bonds to get the new "old road" started.

Although it relates to another highway in Sonoma, Col S. H. Finley, County Highway Engineer of Orange County, wrote the following which can give us a glimpse of how roads were justified at this time:

In your county the savings on depreciation, repairs and fuel for automobiles will average at least $110 each per year. Assuming that you have 2,000 motor vehicles of all kinds in the county, the savings to the owners of these will annually amount to $220,000. Assuming that you have 5,000 horses and mules in your county, the good roads will save at least:

Glenwood post office and service station. The automobile is a Buick. *[UCSC Special Collections]*

- *For shoes and harness annually, $1.50 each* *$7,500*
- *For each animal for feed, 50 cents per month* *$30,000*
- *Increase useful life of each animal, 10%* *$75,000*
- *Wear and tear on wagons and carriages, 2,000 at $5 each* .... *$10,000.*
- *Total savings* *$122,000*

[California Highways, September 9, 1950, p. 77]

Glenwood Highway

The State was asked to recommend a route for this new (paved) highway. Glenwood founder Charles Martin had anticipated this and quickly provided the state with three surveys completed at his own expense. One route was close to the old Santa Cruz-San Jose road ending in Soquel, another followed Bear Creek and the third was the Glenwood route.

Not too surprisingly, the best route on the Santa Cruz County side was through Martin's Glenwood development, which greatly increased the value of his properties. The recommended route followed the railroad bed, which had been constructed 40 years earlier, from Los Gatos, terminating at Ocean Street.

The actual Glenwood Highway route started in Los Gatos on Santa Cruz Avenue, continued south to near Montevina Road, then headed up the center of today's Lexington Reservoir joining the current Old Santa Cruz Highway at the Idylwild intersection. At the summit, the road continued due South, across the current Summit Road, ending up just south of the current summit restaurants. It continued south, crossing 17 and up the driveway that can be seen to the west of 17 across from the Old Santa Cruz Highway intersection. Then up the ridge above and across from inspiration point to where it joined today's Glenwood Drive continuing down the mountain merging with today's Scotts Valley Drive. At this point it widened to 3 lanes, the middle one a passing lane for cars going either direction, and then through Scotts Valley, past Camp Evers, east on Mt. Hermon Road to Glen Canyon Road. Then south until Glen Canyon turns into Branciforte, then turns into Market St. leading to Water Street in Santa Cruz.

San Jose residents could reach the Los Gatos origin point by traveling a route south that generally follows today's Bascom Avenue that turns into Los Gatos Blvd.

Early in the summer of 1912 the grading was started on the new highway, and it was practically completed by 1915. The paving of the first road was started then and the first job finished between Santa Cruz and Scotts Valley. This road carried specifications for pavement 15 feet wide and 4" thick. It extends up to the Sand Hill school house. [Santa Cruz Sentinel, September 5, 1935, p. 4]

Construction was performed by Soss & Marshall, Occidental Construction, Federal Construction and J. L. Connor.

Marshall's workers line up for an official photograph near the Glenwood tunnel, about 1915. *[California Department of Transportation]*

At this time federal funds were secured to aid in financing the highway, which were accompanied by specifications that the road should be 17 feet wide, 7" thick...and be reinforced with steel to give it more resistance to weather. [Santa Cruz Sentinel, September 5, 1935, p. 4]

"I want to see a paved road through these mountains where I first rode on horseback, when there wasn't even a trail," Charles Martin is quoted as saying. Although Martin turned the first furrow making the beginning of the project, unfortunately he died in 1920, at age 83, before completion of the new road.

J.A. Marshall, contractor for much of the work wrote that conditions were rough. In a handwritten letter to the California Highway Commission, dated June 10, 1915, Marshall requests a time extension to the original contract:

The past winter was most severe and we practically accomplished no headway from the middle of December to the beginning of the present month. The rainfall at Wrights was 55.27 inches, Glenwood 58.62 inches and Laurel 61.35 inches. At no time did the line of highway dry out sufficiently to permit hauling even a moderate load over it. Indeed, most of the time it was absolutely impassible for any purpose. .... At several points springs in the fills caused a complete collapse.

Work was primarily done by hand, with some help from mules. Conditions were rough and experienced help difficult to find. There were also humorous moments. Caltrans files contain a handwritten letter by D.F. Duffy of Los Gatos complaining about his boss. In 1921 he writes:

I was working....as a carpenter on the State highway south of Los Gatos.....As soon as the resident engineer (E. Blockley) saw the steel was bent over he commenced bellowing in his usual style and wanted to know "who in the hell did that" and said that the footing was ruined when anyone with good horse sense would have known no damage was done......Finding I was fired, I hunted up Blockley. I told him I was going to settle with him, which I proceed to do, though I guess I didn't give him half he deserved for before I had hardly struck him he began squalling for help like the cowardly cur that he is. When I was tired of punching him ..... I picked up my tools and went home.

The California Highway Commission launched an investigation and sided with Blockley.

While local residents liked the improved road, there were still some complaints. In November of 1915 Santa Cruz County resident Alex M. Locke wrote Highway Engineer A.E. Loder to complain:

I am an old resident of Scotts Valley....Like my neighbors I had great expectation of the State Highway. Under County management our road was always crowned up after every Fall after first rains and every Spring after heavy storms were over, and kept sprinkled all Summer. We had smooth dustless roads during summer and while more or less muddy in winter, hauling could always be done. Since the State took over there has been a sad change. While it is true that grades have been improved and sharp turns most eliminated, of what benefit are these when the surface is allowed to become in such an execrable condition that neither can freighting be done without imminent risk of breakage to vehicles, nor can traveling over it be accomplished without great discomfort.

An early 1920 photograph of the area south of Los Gatos. The old road goes horizontally across the picture. Black Road joints it on the right, Bear Creek Road to the left. *[William A. Wulf Collection]*

Post card showing traffic on the old Glenwood Highway. The view is looking south, coming into Scotts Valley. *[Covello and Covello Photography]*

Color postcard shows the old highway in the Santa Cruz Mountains - with no traffic! *[Bruce Kennedy Collection]*

The deep and fearful chuckholes remind us of the old mill roads of long ago, to which we do not wish to return. ... May I ask what relief we can expect from the present prospect of a repetition of last Winter's experience, when we were almost marooned out here?

Like the later Highway 17 projects, the construction was done in sections through individual contracts. Unfortunately most of the original records have been lost, but Caltrans does have records of contracts 83 and 49 related to work done between Glenwood and the Sand Hill School House (north of today's Scotts Valley). This 3.15 miles of work cost the state $31,372, plus the state supplied $21,258 worth of materials. The final report states that the final concrete pavement was 15-18 feet wide. Traffic was diverted over the Mountain Charlie Road and several temporary dirt roads. According to Caltrans reports: "Detours were under telephone control, installed, operated and maintained by Santa Cruz County." Apparently this meant that flag men were connected by telephone to know when to let traffic proceed.

Janette Howard Wallace, long time Scotts Valley resident, recalls that in Scotts Valley, while the highway was being built, automobiles temporarily had to use the unpaved Bean Creek Road which caused everything within sight to be covered with dust.

The upgraded highway opened to traffic for the first time in November 1915 and the Santa Cruz Mayor Fred R. Howe, as one of the first to try the road, made the 79 mile trip from Santa Cruz to San Francisco in 3 hours, 5 minutes "without changing gears in his Studebaker 6 touring car" [Olin, p. 64].

 A concrete base to the road was not added until 1921-2. Paving was comprised of a 15-17 foot wide 4 1/2 inch deep Portland cement concrete surface having 1 1/2 foot oil treated shoulders. Earth guard rails, mounds of dirt 2 feet high, were created on the shoulders.

A 1915 Santa Cruz Sentinel editorial rejoiced:

Los Gatos is now but a short distance from Santa Cruz by highway. San Jose is now our near neighbor and in fact we will soon be almost a suburb of San Francisco. A weekend trip to this city from the Metropolis will become a common occurrence in the near future. We are just far enough away from the Bay cities to cause their residents to stay here overnight before returning. [Sentinel, December 1, 1915]

Traffic was heavy on the old road. This photograph was taken looking south, just north of Santa Cruz by Sims Road. *[Courtesy of Harriet Ayres]*

Traffic near Alma about 1926. *[William A. Wulf Collection]*

Up until the automobile became popular, visitors normally came to Santa Cruz and stayed several weeks while on vacation. No one at the time understood the impact that the automobile would ultimately have on the city, in which tourists now only wanted to stay for the day instead of for the week.

Another Santa Cruz Sentinel editorial:

If the speed limit of 20 mph is held to, no road could be safer or more enjoyable. The run from San Francisco to Santa Cruz now need occupy but about 3 hours which means that pleasure seekers from that locality may spend the day here and return home in the evening with ease. [*Sentinel*, August 30, 1921]

The so called "Glenwood Highway" was considered the last word in highway engineering. Turns were called "wide" and were steeply banked, in departure from the flat turns of the day. The highway was designed with a 15 feet width, expanding to 17 feet at curves. Much of the road had concrete curbing to help keep people on the road. But if vehicles were forced to go over the curb to avoid an accident or because of a breakdown, the curb drop off often caused the spokes on wheels to break.

The Santa Cruz Chamber of Commerce planned a large caravan and ceremony for August 26, 1921 to celebrate the completion of the new concrete road, but it was never held due to the untimely death of one of the highway's greatest supporters, Santa Cruz Supervisor James A. Harvey. Harvey owned a draying business in Santa Cruz and was killed when a girder which he was unloading for the Soquel Avenue bridge suddenly swung around and struck him on the head.

A 1921 Santa Cruz newspaper proudly proclaims "California's Road of Marvelous Beauty Links Santa Cruz With Her Future Destiny." The new Glenwood Highway is referred to as the finest stretch of highway in California.

Deeper and deeper into the ever increasing beauties of this captivating locality, the concrete ribbon pulls you onward. After glimpsing the interesting settlement at Alma, the ascent over broad, sweeping curves and gentle grades made well-nigh trivial through skillful engineering, is easily negotiable. Registering "ahs" and "ohs" for lovers of scenic beauty, the charming clusters of cottages at Patchen and in the "Valley of the Moon," pass before your fleeting vision through breaks in the trees. And, made a hundred-fold easier than the old, by its widening and super-elevated turns, the last pull to the summit is a joy indeed. [Sawyer]

The Los Gatos Times-Observer reported that traffic along the road had increased 20% from an average of 1,024 cars per hour in 1923 to 1,207 in 1924.

Part of the State's rationale for the new improvements was the military's need to move from San Francisco to Monterey for regular exercises as part of the World War I "Coast Defense Measure." In fact the paving was thicker than usual to be able to accommodate the heavier vehicles and towed guns. In 1987 local historian Margaret Koch, a fourth generation resident of Glenwood, remembered the old Glenwood Highway well.

This was the first locomotive on standard gauge rails into Los Gatos from San Jose (1895). Nine years later the broad gauge line had been extended to Santa Cruz. This wood burning engine, built in 1886, had a unique diamond smoke stack that was designed to catch cinders from the wood fire and prevent fires along the right-of-way. The headlight used coal oil. [*Los Gatos Library*]

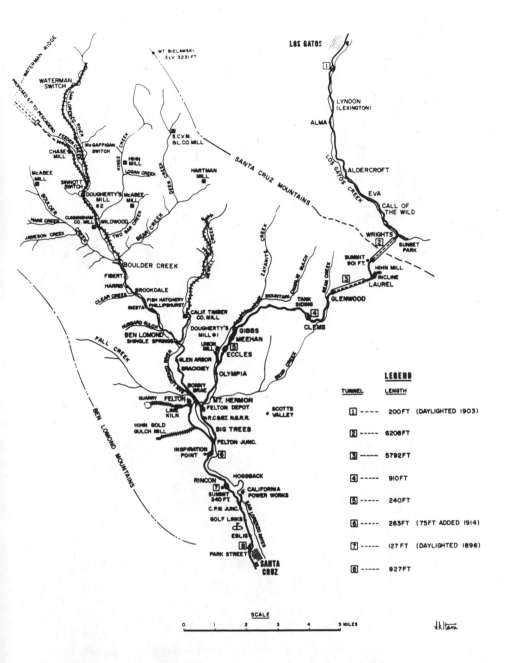

MT BIELAWSKI
ELV 3231 FT

LOS GATOS

WATERMAN RIDGE

WATERMAN
SWITCH

PROPOSED S.P. TO PESCADERO

FEEDER CREEK

McGAFFIGAN
SWITCH

CHASE
MILL

S.C.V.M.
&L.CO. MILL

LYNDON
(LEXINGTON)

ALMA

McABEE
MILL

SINNOTT
SWITCH

HIHN
MILL

LOGAN CREEK

DEER CREEK

KINGS CREEK

HARTMAN
MILL

ALDERCROFT

EVA

DOUGHERTY'S
MILL
#2

McABEE
MILL

WILDWOOD

CUNNINGHAM
CO. MILL

CALL OF
THE WILD

MARE CREEK

BOULDER CREEK

TWO BAR CREEK

BEAR CREEK

LOS GATOS CREEK

JAMESON CREEK

BOULDER CREEK

WRIGHTS

SUNSET
PARK

SUMMIT
901 FT.

FIBERT

HARRIS

BROOKDALE

CLEAR CREEK

FISH HATCHERY
PHILLIPSHURST

NEWELL CREEK

ZAYANTE CREEK

HIHN MILL
INCLINE
LAUREL

FALL CREEK

SHESTA

HUBBARD GULCH

BEN LOMOND
SHINGLE SPRINGS

CALIF. TIMBER
CO. MILL

DOUGHERTY'S
MILL #1

MOUNTAIN

TANK
SIDING

CHIQUINE GULCH

BEAR CREEK

GLENWOOD

GIBBS
MEEHAN

UNION
MILL

ECCLES

CLEMS

GLEN ARBOR

BRACKNEY

OLYMPIA

BEAN CREEK

LEGEND

BONNY
BRAE

BEN LOMOND MOUNTAINS

SAN LORENZO RIVER

QUARRY

FELTON

MT. HERMON

FELTON DEPOT

TUNNEL

LENGTH

LIME
KILN

P.R.C.&ST. N.&R.R.

SCOTTS
VALLEY

1 ---- 200FT (DAYLIGHTED 1903)

HIHN GOLD
GULCH MILL

BIG TREES

FELTON JUNC.

2 ---- 6208FT

INSPIRATION
POINT

3 ---- 5792FT

RINCON

HOGSBACK

CALIFORNIA
POWER WORKS

4 ---- 910FT

SUMMIT
340 FT

C.P.W. JUNC.

SAN LORENZO RIVER

5 ---- 240FT

GOLF LINKS

EBLIS

6 ---- 263FT (75FT ADDED 1914)

PARK STREET

7 ---- 127FT (DAYLIGHTED 1896)

SANTA
CRUZ

8 ---- 927FT

SCALE

0   1   2   3   4   5 MILES

Vk Itani

Vic Itani prepared this excellent map showing the original train routes and tunnel locations in the Santa Cruz Mountains. *[Courtesy of Rick Hamman and Vic Itani]*

A narrow gauge train makes its way up the mountain south of Los Gatos. The old road is seen to the left of the tracks. *[William A. Wulf Collection]*

It was called the Military Highway because soldiers used to come from the Presidio in San Francisco to Monterey on maneuvers....The two lane Military Highway was so jammed with tourists on weekends it got so it was just impossible to go anywhere on weekends because the cars were lined up bumper-to-bumper, overheating. [Bergstrom, pE4.]

Some things haven't changed!

Alongside the road were a variety of small businesses supporting the automobile trade. For example, near the current Sims Road intersection was Mitchell's campground, an old gas station that sold a few groceries and a few small tourist cabins.

One of the early Scotts Valley residents, Ruby V. Strong, lived on the Glenwood Highway near what is today's Malone's Restaurant. She recalls:

Originally there was only a dirt road with an occasional buggy or horse. There was a lot of dust when people would pass by. We would take a wagon and go visit my mother's sister in Los Altos and it would take one whole day. I'd usually ride my horse. I remember the first automobile I saw on the road, before it was paved, was a Model T that someone in Santa Cruz owned. A young man brought his father and they went north up the road a ways. Then the State paved the road and soon there were lots of cars. People loved the new road because it cut down on the dust. There were almost no businesses alongside the road at the beginning but before long small fruit stands and such began to appear. Many of the local men worked on the road.

Railroads

When James G. Fair of the Comstock Lode fame decided to invest some of his earnings from Nevada's silver mines, he turned to railroads. His first project was a narrow gauge railway started in May 1876 from the Oakland area to Santa Cruz, built under the supervision of Alfred E. Davis, President. The initial capitalization was $1,000,000 put up jointly by Davis and Fair.

The inter-continental railroad providing America with its first national transportation system had been completed in 1869, and merchants quickly recognized the large impact that railroads were to have on business and in attracting new populations.

Originally the train route was to have been from Saratoga up to the Waterman's Gap area (junction of Big Basin Road and Highway 9), then along the San Lorenzo River through Boulder Creek connecting with existing tracks at Felton. This proved less practical and more expensive than the eventually selected Los Gatos route. Former San Jose Mercury reporter and writer John Young notes in his book *Ghost Towns of the Santa Cruz Mountains* how different our lives would be today if the original site had been selected and the Los Gatos corridor didn't exist. The San Lorenzo Valley would have taken on more importance, while Los Gatos and Santa Cruz would have been more remote.

The *Call of the Wild* station on the train line, north of the summit. If a flag was put out, the train would stop for passengers, mail or freight. *[William A. Wulf Collection]*

First built by the South Pacific Coast Railroad in 1876, the line was later acquired by the Southern Pacific Railway in 1880. Starting in Alameda, the 3' narrow gauge railway line used extensive tunneling to cross the rugged mountains, and cost up to $110,500 per mile to build. Work was primarily done by several thousand Chinese laborers. The tunneling was especially dangerous and more than 65 men were killed during the work. Tunnel number 1 at the summit took two years to build and cost 30 lives alone.

As the train construction progressed south, towns like Wright's suddenly prospered with business and the future seemed very bright.

Total construction cost estimates ran to $12 million but the railroad was profitable. There were 2 passenger trains and 6-7 freight trains a day, sometimes more on weekends and holidays.

The original train tracks for the Los Gatos route ran from Santa Cruz to Felton, and then east towards Eccles, Gibbs and Glenwood, under the present Highway 17 to Laurel, and then up towards the summit, moving into Santa Clara County to the town of Wright's, then down through Call Of The Wild, Eva, Aldercroft, Alma, and then moved west of Lexington and then to Los Gatos, Campbell, Santa Clara and beyond. These stops were selected because they mirrored the locations of lumber mills.

After nearly 13 years of work the trains were ready to run:

"Betsy Jane" was the first narrow gauge steam locomotive to operate in the Santa Cruz area. [*Covello and Covello Photography*]

The Santa Cruz railroad station about 1890. [*Covello and Covello Photography*]

The South Pacific Coast Railroad opened its completed line on May 15, 1880, a special train being operated over the system from Alameda to Santa Cruz with a picnic being spread at Big trees, just north of Santa Cruz. Tragedy struck at the gay assembly, however, when the train jumped the track and spilled out the celebrants, killing fourteen and canceling the entire opening day program. [Clark, p. 351]

Ten Baldwin, 4-4-0 22-ton steam engines powered the entire line. Freight hauling of lumber and moving fruit from local growers became major enterprises for the railroad.

A standard gauge rail was added starting in 1898, allowing bigger more powerful engines to haul wider cars. The conversion took over 10 years and consisted of adding a third wider rail along side the existing two rails allowing both standard and narrow gauge engines to use the line. On May 29, 1909 the first wide gauge train, train number 84, made the run from Los Gatos to Santa Cruz.

By 1905 there were 25 steam locomotives, 600 freight cars and 75 passenger cars in use.

Six tunnels were built during the initial construction: number 1 running from Wrights to Laurel, and number 2 from Laurel to Glenwood. Tunnel 3 is near Glenwood, 4 north of Eccles, 5 south of Big Trees and 6 near Mission Hill in Santa Cruz. Tunnel number 1 was opened on May 15, 1880 and is over a mile long. It actually crosses the ridge of the Santa Cruz Mountains at an elevation of 903 feet.

Unfortunately, it also crossed the San Andreas fault line and in 1906 was damaged by the San Francisco earthquake. At the location of tunnel 1 the entire mountain moved five feet northwest. Damage was extensive but repairs began immediately and officials used the time to widen the tunnel and install standard gauge track. Other repairs delayed the opening of the complete train route until May 29, 1909.

When the trains stopped running in the 1940s, tunnel 2 was used as a water storage reservoir. Tunnel 4, a shorter 240' run near Felton, was turned into a long term storage vault for important papers (Santa Cruz County and Lockheed use it for storage). Filesafe is the current owner. Several other tunnels near Felton are still in use.

By 1915 the automobile was beginning to make a major impact and train operations began to decline. But until 1940 the train was still the most popular way to travel from the Santa Clara area to Santa Cruz because of the beautiful scenery. It ran daily and was popularly known as the "Picnic Line," although the official marketing name used by Southern Pacific was the "Suntan Specials."

Regular train service was discontinued in 1940 after a severe winter caused great damage to the tracks and the newly opened Highway 17 made automobile travel more popular, resulting in declining passenger service.

If you are interested in learning more about trains, I highly recommend Rick Hamman's book *California Central Coast Railways*.

CHAPTER 4

The Highway 17 Construction Project

The Beginning

The upgraded Glenwood Highway completed over the Santa Cruz Mountains in 1921 only increased the traffic. On Sundays and holidays a continuous line of cars filled the entire 25 mile length.

In 1925 Augusta Emily (Schultz) Springman bought a nine acre parcel near the current intersection of Sims Road and Highway 17 in Scotts Valley. The old road was next to their property and her daughter Harriet Springman Ayres remembers:

That was one of the worst curves of the Los Gatos Highway, just about every weekend Daddy and Mr. Otten had to rescue people out of their cars that went off the highway and landed in Otten's pond during the night. Most of those accidents were caused by dense fog or drunk drivers. [Ayres, p. 5]

The Springman property was later purchased by the State of California and the buildings removed, because the new Highway 17 went right through the property.

The increasing popularity of the Big Basin recreational area and the resort activities in Santa Cruz County had made traffic intolerable. In 1927 Mr. H. R. Judah, Chairman of the California Highway Commission, discussed the "old" road:

89

Highway 17 construction begins on the Santa Cruz side of the mountains. [*UCSC Special Collections*]

Even within the past few years, there have been Sunday afternoons when returning traffic from the Santa Cruz area would backup as far as five miles southerly from Los Gatos unable to move, due mainly to the inadequacy of the road. [Santa Cruz Sentinel, February 28, 1971]

L.G. Marshall, a Caltrans engineer, studied the traffic patterns and found a peak of 17,000 using the cars on one holiday. Caltrans proposed building a new road as:

The answer to 16.5 miles of dizzy serpentine road having 284 sharp curves, the equivalent of 36 full circles! [California Highways, September 9, 1950]

Soquel wanted the new road to end in their city, and Caltrans engineers felt that this was the easier route from a construction point of view. But Santa Cruz politicians prevailed and the current route was adopted.

State Highway 5

At the time the Highway 17 project was starting, the highway was still known as State Highway Route No. 5, the Stockton-Oakland-Santa Cruz Highway. The re-naming to Route 17 happened before the road was completed in 1940. (See Appendix for more on the naming.)

Work on a new "ultra-modern" highway began in 1931, construction started in 1934 and ended in 1940, although work continued until 1943. The new route was approximately five miles shorter than the existing road and was immediately popular with residents on both sides of the hill.

In 1934 Col. John Skeggs, Caltrans District Engineer was quoted:

The highway leads to Los Gatos over long tangents through vast orchards [in San Jose] and in departing from the town....winds its way upward ten miles to a pass over the ridge of the Santa Cruz mountain chain. Then it descends for fifteen miles of devious course to Santa Cruz, through forest covered hills, little friendly valleys potted with homes and patches of orchards - every turn in the road a masterpiece of scenery. And from vantage points over the summit [can be seen] breath taking panoramas of the mountains and valleys and the broad sweep of the sea. [Santa Cruz Sentinel, February 28, 1971]

The new highway was initially four lanes through the mountain sections and three lanes through the valleys and flats. Total distance was 20.6 miles, 14.6 of which was four lanes. It took over 9 years to complete, involved 11 master construction contracts and cost over $3 million. One of the contracts was the largest ever issued by the state, covering 6.5 miles of road where 2.5 million cubic yards of dirt were moved.

The replacement road had 46 easy curves and a minimum of 500 foot-radius turns. The road was constructed with no more than a 6% grade (the section just south of today's summit restaurants is 6%). The original Highway 5 was up to 50 feet wide but has been expanded to 75-86 feet today. This contrasts with the "old" road maximum width of only 21 feet. First the road was graded and a

Colonel (ret.) John Skeggs was Chief Engineer for Highway 17 construction. In the early 30s a typographical error listed his first name as "Jno." and it was incorrectly spelled that way in many publications in the years following. [*California Department of Transportation*]

"bituminous macadam" surface of gravel, sand and oil was put down. The road was actually opened at this point. Years later a concrete layer was added.

The new road was located some distance from the old one due to the "highly developed" area full of cabins, cottages and recreational improvements.

The terrain required special building techniques such as stripping the earth blanket to bedrock and replacing it with imported rock base; constructing heavy rock tow-walls for fills; and most extensive trenching and sub-drains. At one section, through the narrow Los Gatos Creek Canyon, the highway was squeezed between the steep slope of the mountain and a railroad at the bottom. In this location, two of the four lanes are carried on a reinforced concrete and steel pile sidehill viaduct for a distance of about 1,000 feet. [California Highways, September 9, 1950]

Highway construction was the responsibility of Col. John Hunt Skeggs, district engineer. A. Vetter was the local engineer on the Santa Cruz portion, while H.C. Darling of San Jose focused on the Santa Clara side.

Skeggs was born in 1882 in Alabama, receiving his formal education at Auburn. His engineering career began at the Pacific Electric Company in Los Angeles where he helped build the Los Angeles Aqueduct. Soon Los Angeles County recruited the bright young engineer, but in 1917 World War I intervened and Skeggs entered the Army as a Captain. After the fighting stopped, then Col. Skeggs started with the state in 1919 and within two years was named the Division IV Engineer.

Many consider his greatest achievements to be the site selection and connecting roads to the Golden Gate and San Francisco-Oakland Bay Bridges. Besides the

Highway 17 project, Skeggs was responsible for many of the early automobile roads in Northern California including work on Highway 101, the Skyline Boulevard, Pacheco Pass, the Waldo Tunnel, the Altamont realignment and widening El Camino Real.

At his 1952 retirement party at the Fairmont Hotel, Skeggs' successor, B.W. Booker, gave him keys to a new automobile so he could properly enjoy the roads that he had helped to build!

Highway 17 was not built as one complete road as we see it today. Rather the road was upgraded in several different stages, some of them simply widening and straightening the existing old road while in other places a complete new road was built. The basic stages, in the order built, were:

- Santa Cruz to Mount Hermon Road
- Vine Hill Road to Inspiration Point
- Inspiration Point to 1.9 miles north of there
- Los Gatos to Inspiration Point area
- Mount Hermon Road to Vine Hill Road
- Lexington Reservoir area
- San Jose to Highway 9

Santa Cruz to Mount Hermon Road

One of the first controversies about the new highway was that a new route had been chosen, bypassing the town of Scotts Valley. Up until that time, the highway had gone along what is today's Scotts Valley Drive.

Scotts Valley businessmen opposed the new route, instead favoring widening the existing three lane road through town. Businesses were primarily concerned about the effect that the bypass would have on their incomes. Caltrans fought to keep the new route, claiming that it only wanted to build a safer freeway. State officials had more power over local interests than today and the new route was quickly finalized. However the portion of the road from Vine Hill Road south to the Mount Hermon Road exit was one of the last built to appease the Scotts Valley merchants.

The first contracts for the work were signed on May 18, 1931 and construction started by September. The contract called for a new straightened road from the area near today's Highway 1-Highway 17 intersection north past Pasatiempo for a total length of about 3 miles to the area near Mount Hermon Road.

As the rest of the new road became usable, Scotts Valley then complained about the dangerous exit onto the old road. Coming south from San Jose, the new road ended about where the Gramite Creek exit is today and went west to join the old road in Scotts Valley. Where the new road joined the smaller old road there

93

An aerial view of Highway 17 construction from Santa Cruz to Scotts Valley, looking north. *[Covello and Covello Photography]*

Mechanized construction equipment was just coming into use - and greatly increased productivity. *[California Department of Transportation]*

Charles H. Purcell was the State Highway Engineer from 1928-42 during the Highway 17 construction. Later he was appointed Director of Public Works from 1943-51. *[California Department of Transportation]*

was a dangerous curve that caused huge backups. Caltrans refused to extend the road further at that time, citing the prior agreement to build this section last.

The September 22,1933 dedication of the major Santa Cruz portion of Highway 17. Held at Inspiration Point, officials are about to cut the ribbon officially opening the portion of the new road leading to Santa Cruz. Director Earl Lee Kelly is about to cut the ribbon, flanked by officials, bathing beauties and women in period costumes for the upcoming Santa Cruz birthday party. From left to right the men are State Highway Commissioner T.A. Reardon; Caltrans District Engineer John Skeggs; Santa Cruz Mayor Ray Hammond; Earl Lee Kelly; Chairman Harry Hopkins of the Highway Commission and Commissioners F.A. Tetley and P.A. Stanton. Unfortunately the names of the women were not recorded. [California Department of Transportation]

1936 photograph of a three lane portion of the road. Cars travelling in both directions shared the middle passing lane, making head-on collisions inevitable. [California Department of Transportation]

Vine Hill Road to Inspiration Point

This section of the construction was extensive. A new route from the summit area straightened the road considerably, but this caused it to be more costly than the Los Gatos side due to extensive grading requirements.

Work was done by the Mittry Brothers, contractors, at a cost of $618,000 for 6.67 miles of work. More than 1,500,000 cubic yard of dirt were moved. The number of curves in this section was reduced from 130 to 22 and 6,101 degrees of curve, or nearly the equivalent of 17 full circles, was eliminated. The maximum radius of the new curves was 500 feet vs. the old 80 feet. The width was a minimum of 46 feet through the mountains, 36 feet in the valley - about twice as wide as the original road. Some portions were up to 50 feet wide. Driving times were reduced 15-20 minutes because of the new improvements on this section alone.

Power trucks and specialized tools were just beginning to be widely used by construction companies when work began on the road. One hundred and thirty five men, working three six-hour shifts, completed the work.

Ten huge six-yard trucks, recently moved from the big government job at Boulder Dam, are being used in tearing through the mountains. Three huge steam shovels, two Diesels and one gas, each of 2.5 yard capacity are on the job, while there are ten tractors from 30 to 65 horsepower capacity, aiding in the work. [San Jose Mercury, March 16, 1931]

This portion of the road was dedicated on Saturday September 22,1933. Members of the State Highway Commission were fed a noon lunch at the Palomar in Santa Cruz which "was without speech making." Earl Lee Kelly, California Director of Public Works and Harry Hopkins, chairman of the State Highway Commission were the honored guests.

At 1 p.m. they were taken north over the old route by Capt. Jack Payton of the CHP to Inspiration Point where the group stopped for speeches before a crowd of over 1,000. Col. John Skeggs, District Engineer in charge of the project, represented State Highway Engineer C. H. Purcell. Purcell (1883-1951) was famous for being the Chief Engineer for the San Francisco-Oakland Bay Bridge that was completed in 1936, and in 1943 became California's Director of Public Works. Chairman Harry A. Hopkins of the California Highway Commission, and Commissioners Timothy A. Reardon, Philip A. Stanton and Frank A. Tetley were also speakers along with Santa Cruz Mayor Roy Hammond and Fred McPherson, Jr., Master of Ceremonies.

A temporary barrier, begonia-studded with redwood tree needles, was formally broken by Skeggs and Kelly at 2 p.m., and Kelly's official car was driven through the gates to mark the official opening of the new highway. Then the caravan continued back over the new Highway 17 for an evening "preview" of the Santa Cruz birthday party to be held the following month. The party included residents in old Spanish costumes and a bevy of bathing girls. Later the dignitaries celebrated with a dinner at the Resetar Hotel in Watsonville.

Three weeks later a separate paving contract for $179,222 was given to the Hahrahan company of San Francisco after the road was opened to the public on October 10, 1933.

Inspiration Point Area

Another project concentrated on the area from 1.9 miles north of Inspiration Point to Inspiration Point. Again a new route was created to straighten the road.

Union Paving company did the work for $67,484. Union engineers said, "it was one of the 'biggest little jobs' in highway history because of the difficult terrain encountered in the region." [San Jose Mercury, January 14, 1934].

Landslides have always plagued the highway. In February 1935, before the entire road had been completed, local papers reported that:

Every available highway maintenance man - some 40 in all - rushed to Inspiration point to fight the slide of 50,000 cubic yards of earth that has dropped out of the new Los Gatos-Santa Cruz state highway. Seepage from underground spring has been blamed for the loosening of the gigantic fill, which is slowly grinding down the side of the mountain, toppling redwood trees in its path. [San Jose Mercury, February 26, 1935]

At first two lanes were blocked, but the slide continued and soon a third lane dropped 50 feet below the grade. Redwood trees 100 feet tall were toppled and over 50% of the dirt in the area slid. Engineers dug test holes at various locations 35' into the ground to locate the water source - and all of them filled up with water!

The solution was to dig a drainage ditch 30 feet deep, 12 feet wide and 150 feet long to carry the water off in a new direction. Tons of rocks had to be imported to fill the ditch. A temporary detour was created, but the turn was too sharp for trucks and buses which were forced to use the old road. Repairs continued around the clock and eventually cost $30,000. A permanent solution wasn't found until later that year and involved underground piping to divert the water.

The newly completed Inspiration Point turnout. Notice how few trees are in the area. [*California Department of Transportation*]

Inspiration Point slip-out photograph from a official Caltrans incident report dated June 11, 1936. *[California Department of Transportation]*

To make matters worse, tragedy struck and John Kent, 32, of Santa Cruz was killed when the drainage ditch caved in. A coroner's jury, meeting in the Santa Cruz mortuary, ruled that the State Highway Department was guilty of "gross negligence."

Los Gatos to Inspiration Point Area

This section of the construction went from Main Street in Los Gatos to a junction point 1.9 miles north of Inspiration Point.

The major section completed the realignment efforts and was again aimed primarily at straightening the road. It took off south from just south of Oaks Street and cut an entirely new path to Inspiration Point.

Originally the realignment project went only as far as the southern Los Gatos city limits, known by local residents as Windy Point. But city officials were able to talk Caltrans into upgrading a .64 mile two lane section into three lanes between there and the Main Street-Santa Cruz Avenue intersection by offering to pay $7,400 of the construction costs. There was concern voiced by the City Council that a new bottleneck would be created at the Main Street intersection, which would require more city money to fix, and this certainly came true.

Alma and Holy City residents voiced many complaints about the proposed route through the Redwood Estates area. Many complained that the new route would ruin their quiet vacation properties, that "millions" had been spent on road development of the existing road and that the existing towns and busi-

A sheep's-foot roller helps settle the dirt at the Moody Gulch fill. Caltrans construction report photograph dated September 21, 1938. *[California Department of Transportation]*

nesses on the old road would suffer. But eventually the state exercised eminent domain rights and acquired property from 362 individuals and companies including such famous names as the Martins and McKiernans. Total 1937 value of all the land purchased by the State amounted to $57,142.

This section shorted the road length by 1.96 miles, reduced the number of curves from 132 to 20, reduced the turn degrees from 7,700 to 1,118, and increased the radius of curves from 80 to 500 feet. The road was widened to 46 feet for 4 lanes.

Contract for the 6.25 mile section went to a combination of three companies who jointly bid $895,045 for the project: Heafey-Moore, Fredrickson-Watson and the Fredrickson Construction companies of Oakland. A.M. Walsh was the Resident Engineer for the State Division of Highways.

Preliminary work started in January 1938 and took over a year to complete. As many as 260 men were working on the project at the height. 2,200,000 cubic yards of dirt had to be moved and at least two parallel major fault planes were crossed. Costs eventually escalated to over $180,000 per mile due to the difficult terrain. 114 acres of redwood timber had to be logged and clearing costs alone ran to $541 per acre.

Originally a huge rainbow bridge was planned over Moody Gulch but this was dropped due to high costs and instead the area was filled in with dirt (site of todays Moody Curve). Caltrans officials calculated that the bridge alone would have cost two-thirds of the entire project's funding. The change required making the route one fourth mile longer than originally planned and required covering some of the abandoned oil wells in the area.

Watering the embankment to keep dust down during construction in 1931. Notice that mules were being still being used for some hauling despite the availability of automobiles. *[California Department of Transportation]*

Heavy reinforced concrete arch culverts were constructed under high fills where the required waterway was over seven square feet. One such culvert, built under an old arch bridge, is shown above. Both were later buried in dirt fill so the road could cross above the area. *[California Department of Transportation]*

Extensive deep trenching for rock filling in this drainage system was necessary to insure stability of the large dirt fill (1937 official construction report photo). *[California Department of Transportation]*

Our plans are drawn, however, with the consideration that a bridge may be built there in future years, stated Col. John Skeggs, district engineer. *[San Jose Mercury*, September 22, 1937]

Eventually seven concrete arch culverts were installed at Moody Gulch and Aldercroft , Lyndon, Black, Fremont creeks.

On May 22, 1938 Governor Frank Merriam commemorated the official completion of this section of the new Highway 17 using Santa Clara County prune juice in place of champagne. Clad in a brightly colored bathing suit, Gloria Daily as "Miss Santa Cruz" managed in two tries to break the heavy bottle of syrupy prune juice against the iron shovel while a crowd of 500 looked on. The Los Gatos high school band, uniformed in orange and black, started the program with a concert and closed it with the Star Spangled Banner.

At the ceremony held near the site of today's Alma Fire station, the Governor in the principal address praised the road as, "probably the most important recreational road in California." [San Jose Mercury, May 22, 1938].

Mount Hermon Road to Vine Hill Road

Under pressure from Scotts Valley residents and the Santa Cruz Board of Supervisors to fix the bottleneck curve where the new road was diverted into Scotts Valley, construction for this 3.5 mile portion of the highway through Scotts Valley was started in 1936.

Costs estimates were $92,000 per mile but by 1938 had nearly doubled to $180,000 per mile because of the difficulties encountered. Over 2,200,000 cubic yards of earth were moved and 14 acres of redwood trees logged. Drainage and

The dedication ceremonies at Alma Fire Station celebrating completion of the Los Gatos to Inspiration Point portion of Highway 17, held in 1938. *[William A. Wulf Collection]*

seepage caused severe problems making building a stable roadbed difficult. The road width was from 36-46 feet wide and this section was eventually completed two years after it started.

Lexington Reservoir area

From the reservoir north to Los Gatos the road narrowed and caused huge traffic jams. In 1940 Caltrans started a new "sidehill viaduct" project that widened the road by driving pile supports out into the banks of Los Gatos Creek. The project was completed in August 1940, Heafey Moore Company and Fredrickson Watson Contractors did the work. It was especially difficult because of the nearby railroad tracks.

Major traffic backups occurred throughout the construction as the road was forced to narrow into two lanes.

Highway 17 construction just south of Los Gatos. *[UCSC Special Collections]*

Widening the road south of Los Gatos for the "sidehill viaduct" project required driving steel "H" piles with a crane driver and forming pile bent caps before pouring reinforced concrete caps. *[California Department of Transportation]*

Widening the road in the narrow canyon just south of Los Gatos was very difficult due to the proximity of the railroad tracks and the steep bank. *[California Department of Transportation]*

San Jose to Los Gatos

This section of the road wasn't completed for another 20 years. Until 1959, Highway 17 only went from Santa Cruz to Los Gatos. Motorists then went onto Santa Cruz Avenue, east on Main St. which turns into Bascom Ave. into San Jose.

First on the improvement agenda was the Los Gatos "by-pass," completed in 1956. This work created a new road from just south of Los Gatos , along the Los Gatos Creek, as far as Highway 9 (Saratoga Road).

In 1957 a series of contracts for the Los Gatos "link" totaling $5.92 million was awarded to build a new 8.8 mile 4 lane freeway west of Bascom Avenue, from Highway 9 to Bascom Ave. Construction started on July 1, 1957 and was finished on April 30, 1959.

The work involved building 20 bridges, overpasses and other freeway structures. San Jose's O'Connor Hospital protested the new road because it eliminated an exit at Forest Avenue. They wanted an off-ramp there to provide direct access to the hospital, but Caltrans refused to redesign the intersection. Construction began on July 1, 1957 and the new road was dedicated on April 30, 1959 at a ceremony held at the Los Gatos Lodge.

The remaining one and a half miles of freeway between Highway 101 and Bascom Avenue had completed the design phase, but funding was several years away. San Jose politicians wanted the work done immediately, Palo Alto wanted 101 extended further south instead. Palo Alto won the funding war and this section wasn't completed until much later.

Construction of the main portion of Highway 17 was completed in 1940. The new road featured four lanes with the wide sweeping curves. This photograph was taken before traffic lines were painted. *[California Department of Transportation]*

Highway 17 Completed

In August of 1940 there was a 3-day celebration in Los Gatos designated as "trail days" to celebrate the near completion of the Los Gatos-Santa Cruz Highway.

To the motoring public, this is more than a local celebration. Every tourist who has gone through the tortures of seasickness on the old winding road will welcome the new four lane highway, one of the nicest mountain thoroughfares in the U.S. This boulevard has reduced the Los Gatos-Santa Cruz run a full half hour. [Mercury Herald, August 7, 1940]

On August 30th the official opening was held in Los Gatos. Dedication ceremonies featured a parade of old and modern modes of transportation, followed by a luncheon. President Stanley Mills of the Los Gatos Chamber of Commerce introduced officials including Col. John H. Skeggs, District Engineer representing State Highway Engineer C. H. Purcell as well as Deputy District Director of Public Works Morgan Keaton representing the Governor. Much of Keaton's speech was dedicated to arguments for a higher gasoline tax to:

Confront an increasing volume of traffic which necessitates and demands even higher standards of safety and convenience in construction of highway facilities. [California Highways, September 1940, p. 11.]

Skeggs added "I don't think we have ever had a more difficult job to execute."

A treasured possession of the Los Gatos Chamber exhibited [at the event] by Manager W. W. Clark is a stock certificate dated San Jose May 20, 1863, for 66 shares in the Santa Cruz Gap Turnpike Joint Stock Company incorporated in 1857 with a capital stock of $21,000 to build a turnpike road over the mountains. [California Highways, September 1940, p. 11.]

First traffic counts showed 9,000 vehicles a day using the new highway. By the August of 1940 over $1,750,000 had been spent on the roads over the Santa Cruz Mountains and finally residents believed they had an adequate road.

The terrain changes made by the project combined with subterranean springs and wet weather of the region made mud slides a common problem along the route. The winter of 1941-2 was especially severe and for almost a year caused traffic to be reduced to two lanes, one in each direction, at Inspiration Point. There were actually two washouts, about a half mile apart, where two lanes of the new highway and the entire old highway had collapsed. Engineers installed more drains and filled the area with sandy soil instead of clay, in an attempt to make the road bed more stable. In the end, 96 horizontal pipe drains were installed to divert the natural water. The washouts cost the state $64,000 to

Official group shot at the 1940 highway dedication in Los Gatos. Left to right: President Stanley Mills, Los Gatos Chamber of Commerce; Morgan Keaton, Deputy Director of Public Works; Col. John Skeggs, District Engineer; G.A. Morgan, Chairman, Santa Cruz County Supervisors; Supervisor C.P. Cooley of Santa Clara; C.D. Hinkle, Mayor of Santa Cruz; Marc Vertin, City Council Member and Acting Mayor of Los Gatos. The little girls are Nadyne Rhinelander and Cecelia Miller of Los Gatos. *[California Department of Transportation]*

107

Two lanes of the new Highway 17 are temporarily opened while repairs are being made. The other two lanes slipped down the hillside during winter mud slides (1941). *[California Department of Transportation]*

Completed repair of slip-out at the junction of the old and new highways, 12 miles north of Santa Cruz (1941 photograph). *[California Department of Transportation]*

repair. Despite the problems, traffic counts were 4,000 daily, 16,000 per day on weekends. The resulting traffic delays were horrible.

The traffic bottleneck made everyone upset and in June 1941 the San Jose Mercury reported:

J.W. Vickery, state safety engineer, said yesterday that additional warning signs, reflectors, and a flagman as needed, will be posted at the slide near the highway summit where there have been five major accidents since March, killing a woman and maiming two persons for life, and injuring 10. [San Jose Mercury, June 28, 1941]

On February 13, 1942, another big slide closed the road at Vine Hill. The slide was 200 feet across, 30 feet high and contained 80,000 cubic yards of earth. A bus from Santa Cruz carried the mail up to the slide, and the bags were hand carried over the dirt to a waiting postal bus on the Los Gatos side so the mail could go through.

Later Improvements

July 4th weekend in 1950 brought 20,000 cars over the highway each day to the beach.

Traffic was not always bad, however. At night and during daylight off hours the traffic was quite light. Santa Cruz resident Mary Jane Rodriguez remembers that teenagers used to race over the road, utilizing all four lanes (this was before the center divider was installed).

Caltrans became caught between safety and growth controversies. Continuing to widen and straighten the road to provide for more capacity became extremely controversial to the growth-conscious Santa Cruz County residents, so only small safety projects were attempted.

In 1951 a 1.75 mile section of the road south of Los Gatos was re-routed as part of the Lexington Reservoir project at a cost of $1,392,744. Work was done by Guy F. Atkinson Co. of San Francisco. The road was moved west about 300 yards to move out of the newly created lake.

In August 1958 the Highway 17-Highway 1 intersection was completed at a cost of $1,675,879 by San Jose contractors Dan Caputo and Edward Keeble. This eliminated the need to enter Santa Cruz city itself and eliminated the infamous "slaughter house curve." The new intersection diverted visitors in several directions eliminating a huge bottleneck in the summertime.

In 1959 bids were solicited for a 3.3 mile four-lane expressway between two-tenths of a mile north of Santa Cruz to just south of Glen Canyon Road. Estimates were $1.88 million.

Construction is nearly completed on the Highway 1-Highway 17 intersection. The 1958 improvement greatly helped traffic flow into Santa Cruz, eliminating a major bottleneck. *[UCSC Special Collections]*

The Revolt Starts, Improvements Continue

The early 60s mark the beginning of the road revolution. Until this time voters and politicians welcomed new highways and did anything they could to get roads and improvements funded at the state level.

In 1962 Alan Pattee (R-Salinas) introduced a bill to curb the powers of the California State Highway Committee. Giving local communities more power over route adoption was the issue, but it evolved into a much wider discussion about the impact of roads on the quality of life in California. Los Angeles led the revolt but other counties soon followed.

State Highway Commissioner Joseph Houghteling complained :

We have built a perpetual motion machine....All state taxes go to highways. Money from this source cannot be used for other transportation. A certain amount must be spent in even the most isolated counties whether it is needed or not. The inmates are running the asylum! [San Jose Mercury, May 26, 1966]

Highway 17 soon became embroiled in the debate. There were (and are still) two basic themes: improving safety and raising traffic capacity. In 1967 Alan Shart, Caltrans Engineer, said, "The present four lane highway just about reaches its design capacity with an average daily traffic load of 20,000 vehicles, 30,000 on weekends." [Santa Cruz Sentinel, January 12, 1967].

110

But everyone was more concerned about the safety issues. Highway 17 has been labelled at various times as "Killer 17" and "Blood Alley".

In the mid-60s Philanthropist Harvey West (whose gift of Harvey West Park in Santa Cruz bears his name today) had gory billboards put up, with coffins and skeletons in dayglo red, to scare motorists into driving slower. He also paid for fake paper and wood highway patrol cars. On July 4, 1964 he organized pickets advising motorists that death awaited them on the highway.

Instead of focusing on the safety issues, politicians immediately took sides whether the signs were legal or not and one night they were mysteriously chopped down.

The negative publicity had the desired effect, however, and the state jumped the priority for a new Highway 17 freeway up the list. By 1965 they were looking for both new alternate routes and increased funding for safety improvements.

Guard rails were added at Laurel curve, new speed limit signs were installed, a left turn lane created at Vine Hill, a $4,500 left turn lane at Laurel, and a $50,000 barrier north of the summit were all added within a few short years.

In April 1966 a third lane was added southbound between 280 and Hamilton, with a third lane added northbound in May. The contractor was Fisk, Firenze and McLeans of San Carlos.

On July 21, 1966, at the first meeting of the Los Gatos based CRASH group (Citizens Recommending Action Seventeen Highway) John Burgers, spokesman for the group said:

Civic groups, service clubs, neighborhood associations and private citizens have put themselves on record as wanting to do whatever they can to improve the sad conditions that exist on Highway 17. [Los Gatos Times, July 21, 1966)

Caltrans responded by issuing bids for a design review to study new routes from Lark Avenue south to the Santa Cruz County line. Also in 1966, after Highway 17 claimed 5 lives within 10 days, State Senator Alfred E. Alquist (D-

Portions of the Old Santa Cruz Highway still exist east of Highway 17. *[Richard A. Beal]*

Color postcard shows almost no traffic on the "new San Jose-Santa Cruz California Highway" (photo from the 1930s). *[William A. Wulf Collection]*

San Jose) proposed putting toll gates on Highway 17, "like a bridge across the mountains." He added,

I realize this proposal would normally not sit well with Californians, who have traditionally enjoyed use of toll free highways. However this is an emergency situation calling for a top speed solution. [Santa Cruz Sentinel, December 16, 1966]

In 1969 Los Gatos requested a study of alternate routes, especially the so called "eastern corridor route" from Almaden Valley, into the Skyland area, terminating in Soquel. Caltrans Engineers Clifford Greene and Robert Crockett reported they had a beautiful route surveyed but it was simply too difficult to build. Cost estimates were $160 million because several tunnels would be required.

Caltrans also looked at two other routes, one to the east near Lexington Dam and another slightly west of the present 17. Both routes were discarded because of the cost of crossing five earthquake fault zones. Costs were estimated at $30-50 million with a construction time of approximately 11 years. The final result was consensus that straightening the present route was the most feasible and cost effective alternative. But even that choice was too expensive so discussion came to a halt.

Safety improvements continued with the road widened slightly from the Patchen Pass to Granite Creek and from Collins Creek to Idylwild in 1969. A major re-paving effort for 5.2 miles of road south of Los Gatos cost $615,000. In 1970 a median strip and left turn lanes were added to permit safer access to the restaurants at the summit. Cost was $215,000 for this 7/10 mile section improvement.

In late 1969 Caltrans put Highway 17 on the list of roads to be upgraded to freeway status allowing up to 100,000 cars a day capacity. The proposed 8 lane freeway was planned to allow for expansion to 14 lanes if needed.

Alan Hart, Caltrans district engineer, said that:

We will use the lowest gap possible through the mountains, but there will be large cuts. [San Jose Mercury, August 27, 1969}

This caused great debate in Santa Cruz and on March 23, 1971 the Board of Supervisors requested by a 3 to 2 vote to have Highway 17 taken off the freeway list because of concerns about increased traffic volumes. At that time there were an average of 28,000 vehicles a day using the road, peaking at 33,000 on summer holidays. The Los Gatos Town Council also disapproved of the widening with Mayor Defreitas saying "it will ruin our canyon and change our town." Assemblyman Frank Murphy Jr. (R-Santa Cruz) introduced a bill in the state legislature exempting Highway 17 from the state's freeway list. The legislature agreed and Ronald Reagan, then Governor, signed the bill.

In March of 1971 State Highway Engineer Bob Crockett showed a Santa Cruz citizens committee a plan to split the present road just north of the Granite Creek exit creating a one way uphill highway. The plan also called for adding a new downhill highway to the west. There were to be three interchanges at Vine Hill, the Glenwood Cutoff and Glenwood Highway near the summit. Each road

1940s postcard showing the new highway. "Throughout California, graceful wide lane highways lead to all the famous points of interest, passing through regions of boundless beauty. Smooth, safe, and always scenic, people all seem to enjoy the pleasures of motoring, for which California is famous." *[UCSC Special Collection]*

The Summit Road overpass, completed in 1971, was extremely difficult to build. Glen Behm was the Engineer in Charge for Caltrans. *[California Department of Transportation]*

would contain 3 lanes, one of which would be for trucks. Crockett said studies indicated the road could not be split on the Santa Clara side due to terrain difficulties, and that instead the roadbed would have to be expanded. The cost was estimated at $77 million. The idea never gained favor.

Median strips with metal guard rails became popular. The first one was installed south of Laurel in 1971 at a cost of $467,000. Also in 1971 anti-skid proofing (coarse asphalic concrete) was put down for about 4 miles north of Patchen Pass to reduce accidents on the downhill section. The intersection at the

Holy City exit was also improved. Similar projects in 1974 put anti-skid grooves from Scotts Valley to 2 miles below the summit and the northbound lanes from Hamilton to 280.

One of the most welcome additions was a bridge at the Summit Road over-crossing of Highway 17. Completed in 1971, it greatly alleviated traffic jams caused by local summit residents trying to get onto the highway. Glen Behm, Caltrans engineer responsible for the work, was commended for the extraordinary job. Caltrans had to design the bridge to match the curve of the road, preserve area trees, build retaining walls and make it earthquake resistant. Construction took two years.

In 1972 the Camden interchange was expanded and the freeway widened to 6 lanes north to the just completed 280 freeway. Campbell officials protested that the new interchange would clog local streets but were unsuccessful in delaying the work. Lewis Jones Construction of San Jose and Leo F. Piazza Paving were responsible work on the $3 million project.

Extensive winter mud slides closed two lanes of the road in 1973 for almost a week. And in 1974 traffic metering signals were installed at the Camden interchange to regulate the number of cars entering the freeway.

1974 saw more median strips south of Los Gatos, and a nearly disastrous winter. Near the Alma Fire Station, underground rainwater runoffs caused the highway to sink nearly 10 inches. Engineers installed culverts to divert the water under the road. Although traffic was re-routed around the east side of the reservoir during the construction, the road was never completely closed. In January 1975 a $740,000 project raised the highway a foot and widened the road to permit a median strip from just south of Alma College Road to a point three-tenths of a mile south of Bear Creek Road.

During January 3-4, 1974 a large snowstorm closed Highway 17, dumping 2 feet of snow in a few hours. More than 100 people were stranded all night in their autos at the Summit. Caltrans later installed "chains required" signs near Black Road and the Santa's Village turnoff so in the future the CHP could turn back cars not equipped for winter weather.

Memorial Day, 1984, was memorable for 3 days of almost total gridlock over the mountains, as people tried to escape hot San Jose for the Santa Cruz beaches.

In 1985 a $17 million project began to re-pave the highway from Los Gatos to San Jose, adding lanes near Hamilton and a concrete safety barrier in places. Construction took almost 2 years to complete.

In July of 1987 $5 million was allocated for an overpass at the Bear and Black Road intersections by Lexington Dam but this is still under review. Discussions about this intersection actually started in 1971. Given current progress, construction could start in the mid-90s.

The 1989 earthquake resulted in many sections of the road being improved, including resurfacing portions, repairing and extending the center divider and guard rails.

115

In 1991 a major re-paving project was started after considerable controversy. Santa Cruz merchants wanted the work delayed well past the scheduled summer start date because they feared it would cause tourists to cancel trips to the area. The work consisted of a $1.9 million re-paving effort from Highway 1 to Highway 9. Some culverts were repaired at the same time at a cost of $1.0 million. Also the Summit Road acceleration and deceleration lanes were widened slightly costing $800,000.

CHAPTER 5

The Future of Highway 17

Introduction

Traffic congestion has increased more than 50% in the past 10 years. Traffic planners say it will increase another 20% by the year 2,000. Gridlocks and long delays are increasingly common.

How long does it take to build a road? Caltrans says that 4-9 years for a less complicated job, 9-14 years if an environmental impact study is required. If there is controversy, as in the Bear Creek interchange proposal, the building cycle can take 25-30 years.

Funding for roads primarily comes from the 31¢ a gallon gasoline tax.

Near Term Improvements

There are no major improvements to Highway 17 scheduled - and only a few minor improvements are funded. The sad reality is that the political process has become gridlocked, with voters knowing that a transportation crisis is coming - but politicians not believing that voters are willing to support both the necessary lifestyle and financial changes needed to fix the problem. Politicians also believe that limiting transportation is a way to slow down area growth.

Looking south from Los Gatos, the newly completed Highway 17 widening makes the road four lanes wide. Traffic lines have not been painted at the time of this photograph, and there are no guardrails. [*California Department of Transportation*]

There are two separate issues involved: safety and growth. Everyone supports making the road safer, but how to pay for it is controversial. Caltrans has repeatedly said over the years the only way to make Highway 17 safer is to upgrade it to a freeway with the resulting straighter wider road, and safety features. But that raises the growth issue for Santa Cruz.

If Santa Cruz wants to have an individual destiny [distinct] from the Santa Clara Valley, then we do not want to tie ourselves even more firmly to the Silicon Valley so that our whole population and their population see that we're just a very quick freeway trip away. Gary Patton (Supervisor, Santa Cruz County), 1991

Carl Williams, the California Transportation Department Assistant Director responds:

A lot of people are being injured and killed on that roadway; there are safety considerations that I think transcend the issue of "no growth". We just can't let people kill themselves out on a dangerous roadway.

Two years ago Highway 17 was designated as an "inter-regional road." This designation means that only the state (Caltrans) has legal authority over it - the counties do not. But political reality is that the state will take no action unless the counties agree. Cecil Smith, Chief Planner for District IV of Caltrans says, "We're not going to force the issue."

Although Rep. Leon Panetta, D-Monterey, agrees Highway 17 is, "clearly a safety hazard to the people to use it," there are only three minor improvements on the official schedule to the year 2000:

1. Add more concrete barriers to 3 sections that don't now have them (construction to start early 1992)

2. At Laurel Road, widen the southbound left turn lane (construction to start summer of 1992)

3. Between Lark Avenue and Camden build a new intersection where the 85 Freeway will intersect (work started in late 1991, completion in 1994).

There has been discussion, but either incomplete funding or continuing public hearings for eight additional proposed changes:

1. Improve the Highway 1–Highway 17 interchange and add a third lane from Highway 1 to Pasatiempo exit ($20 million). This is likely to happen in the late 90s.

2. Add new truck climbing lane northbound near the summit ($3 million).

3. Improve interchange at Granite Creek Road, especially the on-off ramps. Look for changes in the 1998 time frame.

4. Put signals and improve southbound on ramp at Mount Hermon Rd. This is likely to happen the summer of 1992.

5. Widen to 6 lanes from Pasatiempo Dr. to Granite Creek Rd. ($25 million).

6. Add two HOV lanes (one each direction) from Camden to Highway 85 intersection.

7. Add two HOV lanes (one each direction) from Highway 85 to Highway 9 intersection.

8. Create new intersection(s) by Lexington to serve Black, Bear Creek, Old Santa Cruz Highway and Montevina.

Item number 8 has been in discussion for over 20 years, but it finally appears that consensus is near. $6 million has been designated for this future project and if approved, construction will complete the summer of 1994. Only an interim solution is covered and details have not been worked out.

The Santa Cruz County Regional Transportation Commission 1989 report calls only for adding one HOV lane in each direction on Highway 17 between Highway 1 and Granite Creek Blvd. in Scotts Valley by 1995 (Option #5, above). And that is extremely unlikely to happen by then because there has been no planning activity or land acquisition as of the time this book was published.

Caltrans does have preliminary plans to improve the Highway 17-Highway 1 intersection but there is not sufficient funding available for it.

After years of controversy, a new Highway 85 intersection will be added north of the Lark Avenue interchange in the 1994. The Highway 85 extension will be a new freeway section connecting with the present Highway 85 near Cupertino, continuing south towards Los Gatos and then parallel to the mountains to the Almaden area where it will intersect Highway 101 south of the Blossom Hill interchange. Once completed, the new freeway should relieve some of the current traffic congestion experienced at Lark Avenue.

Rail Transit

A mass transit system like BART, a high speed bullet train or even a light rail alternative are not under study, even though this is probably the best long term solution.

No current agency is responsible for developing rail systems. This would require formation of a new "Joint Powers Authority" encompassing both counties. Caltrans recently became responsible for CalTrain but the organization is still "road focused" and has no mandate to take control of statewide rail development. One private group is actively planning to start regular train runs over the mountains.

A monorail alternative was first raised in August 1969 by Santa Cruz County supervisors but nothing came of it.

In 1971 a pilot study done for Lockheed Missile and Space Company by their engineer Alan Goetz proposed rebuilding the old rail line. He found that 37% was still in use, 27% could be easily repaired, 36% required new construction or was unknown. In 1978 Caltrans studied what it would take to rebuild the old railroad tracks between San Jose and Santa Cruz - concluding that it would cost $36.9 million just to restore the rail bed.

In the past, Santa Cruz has shown no interest in rail alternatives. Supervisor Gary Patton is even opposed to studying the idea further:

We did study it. We found that it would cost on the order of $100 million at that time (in the 70s) and that it would have added transportation capacity on the order of a 12-

lane freeway. The question I have is, do we really want to invest $100 million in order to increase our ties to Santa Clara County? [Santa Cruz *Sentinel*, April 29, 1991]

By contrast, Santa Clara County is betting heavily on public transportation for the future - a combination of bus, light rail and BART. Led by Santa Clara Supervisor Rod Diridon, Santa Clara is putting a large portion of their transportation money into rail alternatives. Current plans call for a light rail line corridor from Los Gatos to San Jose, along the basic Highway 17–Vasona route, by the year 2,000. The environmental impact report (EIR) is due by early 1992 and funding is being sought to purchase right-of-way from Southern Pacific. Santa Clara long range planners believe that a light rail system over the mountains to Santa Cruz will ultimately be developed to relieve traffic problems. Supervisor Diridon recently stated:

A rail line over the top is much more environmentally sensitive..at just a fraction of the cost.....I don't want any more people killed on that hill, or more pollution problems. [Santa Cruz *Sentinel*, April 29, 1991]

CalTrain will be developed from San Jose to Salinas by the end of the century and there is some discussion about an eventual line to Watsonville along the 152 route. This is an attractive long term solution for several reasons. Because the land is flat, it is cheaper to build rail beds. Also the population centers for both counties are shifting south so connecting them at the southern point makes sense. And finally, Caltrans is pushing the CalTrain solution.

A private group incorporated in Felton in 1988 under the name Eccles and Eastern Railroad Company, is headed by Rick Hamman. They have 107 stockholders and hope to go public in 1992. They need to raise almost $50 million through a combination of debt and stock. An unusual feature of the corporation

This Boeing-Vertol sketch of a light rail electric train for use over the mountains was done as part of a 1971 study by Alan Goetz. *[Courtesy of Glen Hughes]*

121

is that 10% of all profits will be divided among non-profit historical societies, institutions and research individuals whose work supports transportation history and related subjects.

Their intention is to rebuild the old railroad bed and run seven passenger and two freight trains daily between Santa Cruz and the Los Gatos area. The 14.4 mile run would take 75 minutes. They already have a 133-ton 1904 oil fired Baldwin steam engine, number 2706, that is being restored. It's the same engine that sat on display for years in Ramsey Park, Watsonville.

Although the basic old railroad route will be followed, some changes will be made to straighten the run allowing speeds up to 40 mph. And since the old tracks were covered by the Lexington Reservoir, a new route west of the reservoir and east of Highway 17 will be built including a new tunnel from the northern end of the reservoir into the area where Main Street in Los Gatos crosses Highway 17. At each end, Eccles and Los Gatos, they would then connect to existing track into Santa Cruz and San Jose respectively.

Phase I of their plan goes as far as Glenwood and will feature a special evening train with dinner, dancing and a private cabin, returning to Santa Cruz in time for breakfast.

Although it is not required, Eccles and Eastern are doing an environmental impact report which is certain to generate controversy. If all goes well, the first trains could be running as early as the summer of 1993. The approval process for a train is very unclear and it will be interesting to watch what happens. For more information you can contact them at P.O. Box 818, Ben Lomond, phone (408) 336-3534.

Because of the 1988 California Clean Air Act, government officials are required to place increased emphasis on non-automotive transportation solutions. The Suntan Specials may return!

Toll Road

In 1987 the idea of a 25¢ toll for Highway 17 was quickly killed. Caltrans is strongly opposed to converting any existing roads to toll road status because it violates one of the earliest principles of our road system - that it be available to all citizens. Some local transportation officials believe the idea was dismissed too quickly and that charges for such inter-regional roads could be an appropriate solution.

Long Term Solutions

The Caltrans "route concept" for Highway 17 is a four lane standard freeway with full shoulders, a standard median, including additional slow moving vehicles lanes where appropriate. There are no plans to make this happen.

The 1990 Santa Cruz County Regional Transportation Report states:

The Commission's continuing policy is not to increase the capacity of Highway 17 north of Scotts Valley in support of management of the jobs/housing imbalance.

Laurel Curve in Santa Cruz County above Scotts Valley. The old town of Laurel was 2 miles to the right. *[California Department of Transportation]*

The current thinking among government and private transportation experts is that residents must find new ways to use Highway 17, other than the traditional one-car/one-rider custom. 25% of all travellers are work commuters. If people simply shared a vehicle with one other person, once a week, it would reduce traffic by 20%. Obviously, this postpones, rather than solves the basic problem. Vehicle occupancy on Highway 17 averages 1.22 persons per car in the morning, 1.44 persons per car in the afternoon.

Transportation agencies are encouraging carpooling, van pools, rapid transit, shifting hours, etc. Employers are being urged to take more responsibility in addressing the needs of their employees. In Los Angeles, for example, a recent law popularly called "Article 15" sets a traffic density goal for various size businesses and will assess fines on businesses whose employees don't meet the goal. This shifts the economic burden to the private business sector - especially for large companies. Santa Cruz has formed a non-profit company called the Transportation Management Association (TMA) to help private companies design transportation programs for their employees and represent companies with more than 9,000 employees. For more information on Santa Cruz TMAs, call (408) 423-1111.

Another less popular innovation is called the HOV (high-occupancy vehicle) lane. These so-called "diamond lanes" require that only cars with 2 or more riders can use them during peak hours. Dogs and mannequins don't count but children do. Most county transportation planners say that all future lane additions will be designated as HOV in their attempt to get people to share automobiles.

Regional Transportation Commissions

In the 1970s, it was realized that traffic was a regional, not a local, issue in this day of the two car family and long distance commuting. So Regional Transportation Commissions were created by the Transportation Development Act of 1971 to provide a mechanism for making recommendations to Caltrans about what voters really wanted to have happen regarding both short and long term solutions.

The California Transportation Commission, appointed by the Governor, annually adopts the State Transportation Improvement Program (STIP). The STIP is a five to seven year capital improvement plan which is the *de facto* list of what improvements will happen to state highways. The California Legislature then funds those projects so just getting on the STIP priority list doesn't necessarily mean it will happen - but does give it a high priority.

The STIP, in turn, acts on the basis of recommendations from the Regional Transportation Improvement Program (RTIP), Santa Clara calls their plan the "T2010." Two Regional Transportation Commissions are involved with Highway 17 - Santa Clara and Santa Cruz. And therein lies the basic problem. You will note that the Regional Commissions basically mirror the two county

Highway 17 south of Los Gatos in the 1940s. *[California Department of Transportation]*

political boundaries. And no one group is looking after the Highway 17 corridor. In fact, both the regional plans do not even show the mountain section of Highway 17 on their maps!

Although people often think of Highway 17 as "Santa Cruz's highway," Santa Cruz believes the transportation problems largely come from the high growth rates in Santa Clara County - and that that county hasn't paid its fair share to solve the problem. By permitting new companies to start up in the Valley, without providing housing for their employees, people have been forced to live in Santa Cruz County. Thus Santa Cruz inherited the problem and doesn't have the economic base to solve it. They would like to see increased funding from Santa Clara County.

A decision to widen Highway 17 will fundamentally alter the future of this [Santa Cruz] community. If we add more lanes to the highway, they will be used to capacity and it will destroy the independence and uniqueness of this community. The only thing that gives us [Santa Cruz] any chance of maintaining our quality of life here is that mountain. [Santa Cruz Supervisor Gary Patton, 1984.]

A good indication of the feelings in Santa Cruz is that at the time of the debate whether to designate 17 as a freeway, then Santa Cruz Supervisor Henry Mello, now a State Senator, actually introduced a resolution to make Highway 17 a one-way road northbound from Santa Cruz. It failed on a 3-2 vote.

Santa Clara County voters don't really care about commute problems for Santa Cruz residents - they have their own gridlock issues. The summer beach traffic problem only exists for 5-10 days a year and they believe it is a problem that Santa Cruz should solve if they want the tourism dollars. Los Gatos generally acts as the spokesperson for the County on Highway 17 issues and is opposed to widening the road because it would bring more traffic and more pollution to their area without resulting benefits.

So both counties are opposed to more lanes in the mountains, but will have widened the roads leading to the mountain section to six lanes by the end of the century. They like that the mountain acts as a "meter" to restrict the amount of traffic.

Obviously a solution is needed that satisfies needs in both Counties. Five times as many cars use Highway 17 as in 1948. It is only getting worse.

Santa Cruz simply doesn't want to address the problem because they don't want more residents with jobs over the hill - but new job development is still a controversial issue in the county. Santa Clara will take no action until Santa Cruz shows an interest. Government planners believe that 17 will never be widened - there is too much opposition, the cost is simply too high, and the environmental air quality concerns will prevent any further automotive solutions.

There is some hope about the funding issues. A Bush administration plan announced in February 1991 creates a new category of road - "nationally significant" roads which are different from "federal" roads. Highway 17 is on the nationally significant road list which makes it eligible for 75% matching funds for repairs, improvements and expansion. As a state road, 17 currently qualifies for a 60% matching grant from the U.S. Department of Transportation. The new Bush $87 billion dollar plan to upgrade roads is far from being finalized at this writing - but it will raise the Highway 17 issue once again for the community.

Interestingly, transportation planners in both counties indicated there is very little public concern about Highway 17. Santa Clara residents are much more concerned about traffic on 280, 880 and 101 - and Santa Cruz residents gripe about Highway 1.

If you are concerned about the future traffic density and safety on Highway 17, you should contact the people listed below. They are our representatives on transportation issues - and they have great power over priorities. They want your ideas and concerns, so don't be afraid of contacting them.

Santa Clara

Santa Clara County Transportation Agency
#207, 1555 Berger, San Jose, CA. (408) 299-2884. Larry Reuter, Director

	Representing
Karen Anderson	Saratoga
Jim Beall	San Jose
Ray Bunt	Morgan Hill
Barbara Conant	Campbell
Dave Fadness	Public Member
Jack Going	Public Member
Laura Herrera	Public Member
Manuel Herrera	Public Member
Harry Kallshian	Public Member
Paul Kloecker	Gilroy
Liz Kniss	Palo Alto
Barbara Koppel	Cupertino
Ted Laliotis	Los Altos
Sue Lasher	Santa Clara
Jack Lucas	Monte Sereno
Henry Manayan	Public Member
Frank Maxwell	Public Member
Christopher Moylan	Public Member
Jack Perry	Public Member
Yolanda Reynolds	Public Member
Michael Rothenberg	Public Member
Skip Skyrud	Milpitas
Amal Sinha	Public Member
Art Takahara	Mountain View
Barbara Tryon	Los Altos Hills
Brent Ventura	Los Gatos
Ray Villareal	Public Member
Barbara Waldman	Sunnyvale

Santa Cruz

Santa Cruz County Regional Transportation Commission

701 Ocean Street, Santa Cruz, CA 95060. (408) 425-2788. Linda Wilshusen, Executive Director

	Representing
Jan Beautz	Private Operators
Bart Cavallaro	Santa Cruz Metropolitan Transit District
J.M. Ellis	Caltrans
Ronald Graves	City of Capitola
Fred Keeley	Supervisor, 5th Dist.
Don Lane	Transit
Robley Levy	Supervisor, 2nd Dist.
Todd McFarren	Transit
Joe Miller	Scotts Valley
Gary Patton	Supervisor, 3rd Dist.
Oscar Rios	Transit

CHAPTER 6

Transportation Alternatives

Introduction

If you're like most Americans, the thought of having to share a car with someone is not desirable. But the reality is that we no longer have alternatives. Why not try one of the ideas below? Remember sharing a ride just one day a week would reduced traffic by 20%!

Flexible Hours

The high density hours are from 6-9 am and 4-7 pm weekdays. Can you arrange your travel hours to avoid these?

Telecommuting

Telecommuting is now possible for many workers. Using computers and modems, workers can now work from their home but yet be connected to the central office. Electronic mail, on-line databases, electronic modem file transfer and FAX machines are now common - especially in Northern California. Combined with a conventional telephone, many people can work very effectively at home, at least part of the time. Try to work at home one day a week.

Van Pools

People who regularly commute in vans love them! Vans these days are air conditioned, have large individual seats and reading lights, and are generally quite comfortable.

You can lease your own van and setup your own van pool business fairly easily. The VPSI Computer Vanpool company, a subsidiary of Chrysler Corporation, will lease one of four van configurations from 7 to 15 passengers at a price from

Highway 17 in September 1940, south of Los Gatos. *[UCSC Special Collections]*

$865 to $1,135 per month depending upon the size, equipment and daily round trip mileage. The monthly fee includes the vehicle, insurance, maintenance, 24 hour towing and emergency road service, back-up vehicles, vehicle tax, license and registration. For more information contact VPSI, 12409 Slauson Avenue, Suite J, Whittier, California 90606 voice (213) 945-3983 fax (213) 945-3243.

If you're interested, contact your employer and see if they'll pay the lease costs! If you ride in a van pool which is not administered or subsidized by your employer, you can receive a tax credit equal to 40% of your van pool subscription costs. The maximum annual credit is $480 and your van pool must have 7 or more daily commuters.

For more information call (408) 429-POOL in Santa Cruz County or (408) 996-POOL in Santa Clara County. You also might try the Rides for the Bay Area Commuter at (800) 755-POOL. All of these folks will help you find an existing van pool or help you with getting your own van. They also run training programs for employers who wish to get involved with van pools.

Car Pooling

Car pooling, like van pooling, involves the sharing of private vehicles. Many people do this on an informal basis, splitting the driving days or the cost. They save money this way and have someone to talk to during long commutes. The same agencies that handle van pools can also help you find someone to share private vehicle rides with. Call the "POOL" numbers above for more information.

Park and Ride

Park and Ride lots are areas where you can park your car or bike (normally at no charge) to meet your car pool or van pool or take public transit. First-come, first-served. Both Santa Cruz and Santa Clara counties negotiate with private property owners to establish these lots next to major commute corridors.

Current Locations in Santa Cruz County are:

• Shared Use

1. Resurrection Church, 7600 Soquel, Aptos

2. K-Mart, 2600 41st Avenue, Soquel

3. King's Village Shopping Center, 230 Mount Hermon Road, Scotts Valley

4. Old Sky Park airport (Scotts Valley), go west on Mount Hermon Rd., turn right on Kings Village Road.

- **Informal areas**

1. Freedom Blvd. and Highway 1

2. Residential areas around intersection of Highway 1 and Highway 17

3. Pasatiempo Drive interchange (with Highway 17)

4. Mt. Hermon Road interchange (with Highway 17)

5. Summit Road interchange (with Highway 17)

6. Street in front of Denny's, 6014 Scotts Valley Drive, Scotts Valley

- **Santa Cruz Metropolitan Transportation District lot**

1. Soquel Drive and Paul Sweet Road (near Highway 1). You must pay a monthly fee to use this lot.

In Santa Clara County, park and ride lots exist at:

1. In Los Gatos on the north side of Saratoga Ave. between University and Santa Cruz Blvd. In the public parking area behind the shops.

2. On Winchester in Campbell, the shopping center lot north of Latimer Ave.

3. In San Jose at the south east corner of Branham and Camden.

For more information, contact the:

Santa Cruz County Transportation Commission, 701 Ocean Street, Santa Cruz, (408) 425-2788.

Santa Clara County Transportation Agency, #207, 1555 Berger, San Jose, (408) 287-4210.

17 Express

The 17 Express is a large bus that provides inexpensive transportation between San Jose and Santa Cruz. Scott Galloway, General Manager for Santa Cruz Metro, says that 700-800 commuters use the bus daily. Buses run at least hourly and during commute hours run every 20 minutes. The service is weekdays only - and costs $2.00 one way, or a monthly pass may be purchased for $65. Tickets can be obtained from the Santa Clara County Transit (408) 287-4210 or Santa Cruz Metropolitan Transit District (408) 425-8600. Two buses transit between old Sky Park Airport parking lot in Scotts Valley and 3rd street and San Fernando in San Jose. To get to the Scotts Valley lot, go west on Mount Hermon Rd., turn right on Kings Village Road and follow the signs.

The Highway 17 Express bus has been extremely popular with Santa Cruz riders. *[Richard A. Beal]*

Stops are made at:

Highway 1 and Soquel Road (Paul Sweet Park and Ride Lot) some trips

Pasatiempo/Highway 17 (some trips)

Scotts Valley Park & Ride lot (old Sky Park Airport))

Bird Ave at San Carlos

San Fernando at Montgomery

San Jose Train Depot for Amtrak, CalTrain and Southern Pacific at 65 Cahill

Santa Clara at Cahill

Santa Clara at Delmas

Santa Clara at Guadalupe Parkway

Santa Clara at Almaden Blvd.

Santa Clara at Almaden Ave

Santa Clara at First Street 3rd and San Fernando (one block from San Jose State)

Maybe your employer will pay for your pass!

Taxi

It costs $55-60 to take a taxi over Highway 17, not the cheapest way over the hill, but you can accommodate up to four passengers in the car for this price.

Yellow Cab   (408) 423-1234 Santa Cruz

                (408) 293-1234 San Jose

Buses

Greyhound and Peerless buses run between the 425 Front Street metropolitan transit center in Santa Cruz and Oakland. There are stops in Los Gatos, the San Jose Airport, San Jose, Fremont, Hayward and Oakland.

The Airporter is a private company which runs several vans between Watsonville, Capitola, Santa Cruz, Scotts Valley, the San Jose Airport and the San Francisco Airport. Trips leave about every 2 hours and cost $21-26.

To obtain the latest schedule and stops call:

Greyhound   (408) 423-1800

Peerless   (408) 423-1800

Airporter   (408) 423-1214

CalTrain

If you need to connect to the CalTrain or Amtrak train depot in San Jose, from Santa Cruz, the CalTrain Connector leaves from the Santa Cruz Metro Center 425 Front Street, Santa Cruz, and connects directly to the San Jose Depot. The van leaves approximately every two hours and costs $5.00. There are 9 trips weekdays and 8 trips on weekends and holidays. There are large reclining seats with reading lamps and lots of luggage space.

For more information call (800) 558-8661.

Highway 17 Technical Description

Introduction

Highway 17 is the major arterial corridor between Santa Cruz and Santa Clara County. The route provides the primary access to Santa Cruz residents who work in Santa Clara County, for recreational traffic from Santa Clara and other counties to Santa Cruz beaches, and for trucks carrying goods to and through both areas.

The Name

At the time, Highway 17 construction was first begun in 1934, the new road was called Highway 5 and was part of a path from Santa Cruz to San Jose, up the east bay to Oakland and then across to Stockton. Different portions were designated with letters ("A," "B," etc.) for maintenance purposes. In 1937, Caltrans decided to make highway numbering consistent in the state and re-numbered many of the roads.

The new scheme called for the portion from Oakland to Santa Cruz to be called "Highway 13." The Santa Cruz Chamber of Commerce objected and passed a resolution on November 5, 1934 calling for a different number. Caltrans yielded under pressure and the road was renamed Highway 17.

About this same time Caltrans adopted the roadside mileage markers as the way to identify specific spots on the highway.

In 1978, Los Gatos Historian William A. Wulf proposed Highway 17 be renamed to honor Father Fermin Francisco De Lasuen, who had selected the site for the original Mission Santa Cruz and is credited with opening the road from there to Mission Santa Clara. State Senator Jerry Smith, D-Saratoga, introduced the resolution but Santa Cruz was miffed. "I think it's presumptuous for San Jose to name our highway," said Santa Cruz Supervisor Chairman Ed Borovatz. Political pressure was applied and the idea was dropped.

In 1985, primarily to be eligible for a $100 million pot of federal funding, the Nimitz section of the road from the U.S. Highway 280 intersection through Oakland was renamed U.S. Highway 880 at a cost of $380,000 for new signs.

Size

Highway 17 is 26.37 miles long., 13.82 miles in Santa Clara County and 12.55 miles in Santa Cruz County. Although the road is almost entirely 4 lanes (2 in each direction), northbound the road widens to 3 lanes at Camden Avenue and to 4 at Hamilton - southbound the road starts at 5 lanes at Highway 280, narrows to 3 at Hamilton and finally to 2 at Los Gatos.

The Alma store as it appeared in the 1930s. Highway 17 soon diverted traffic from the old road and the Alma businesses suffered greatly. [William A. Wulf Collection]

136

Type of Construction

The road was built, or has been re-built at different times since its original construction - but was initially made of 8" thick Portland cement concrete . This type of paving is designed to last 25 years but the heavy trucks and extra sideways stress due to the large number of turns in the mountain section require Caltrans to resurface the road more frequently. The type of asphalt used on Highway 17, known as open grade mix, is a porous mixture designed to prevent ice buildup by letting water filter through. It must be replaced every 5-7 years. Resurfacing adds a 1" cap of asphalt concrete.

Some portions of the road have skid reducing groves cut 1" apart across the road to provide better traction.

Expressway vs. Freeway

Freeways are our best highways. The road path is generally straight, there is more capacity, curves are gradual, there are wide shoulders, on and off access points are limited and controlled. You can only stop on a freeway if there is an emergency; pedestrians and bicycles are forbidden. As a result, they are our safest roads.

An expressway, by contrast, is less safe. As on portions of Highway 17, driveways directly enter the road, lanes are narrower, and stopping at turnouts is permitted.

Highway 17 is a 4 lane divided freeway from the 280 intersection to just south of Los Gatos, north of the Cats, and then it becomes a 4 lane expressway over the mountains until entering Scotts Valley at the Granite Creek turnoff where it once again becomes 4 lane divided freeway. There is an additional small section of road on either side of Sims Road in Santa Cruz (about 500 yards in either direction) that is also expressway.

The expressway portion of Highway 17 (in the hills) is dangerous for another reason. Many of the curves are not a constant radius and the grading of the road is sometimes off camber.

On freeways, curves are designed in a way that permits the driver to experience the curve as a constant angle. Normally you can set the steering wheel when you first enter the curve and leave it there all through the curve. This gives a safe feeling of control because there are no surprises. By contrast, several of the curves on Highway 17 are decreasing radius curves while others have two apex points. This means that as you drive through the curve, you need to turn the steering wheel tighter and tighter. Most of today's drivers are used to constant radius turns so they are surprised when in the middle of the curve they must turn the wheel more. Drivers often experience mild panic in this case and turn the wheel too far, pulling the car into the adjacent lane or the barrier. Drivers not used to the road are especially vulnerable to this danger.

Drainage is a potential problem on Highway 17 and some of the roads are slanted to insure proper water runoff. We are used to roads that slant up on the outside of curves - but 17 has several curves that do the opposite. Again, this is a very unsettling feeling and causes drivers to feel they are going too fast for the curve. Erratic driving is the result. What should you do if you are not familiar with the road? Drive slower than normal and you'll do just fine!

Speed Limits

The original speed limit for Highway 17 was 45 miles an hour but it was later raised to 65. In response to safety concerns, it was lowered to 55 along the Santa Clara County side. In July 1960 it was lowered to 50 in the mountain expressway section. The freeway sections today are 55 m.p.h..

Traffic routinely moves at 60-65 m.p.h. during commute hours and when the traffic is light. Regular commuters feel comfortable at this speed except for a few curves where 55 is more appropriate. But those not familiar with the road generally feel safe at 50 m.p.h..

The speed limit for empty or loaded trucks on downhill sections is reduced to 35 m.p.h.. Northbound, the truck reduced speed limit extends from the summit

Although the official speed limit is 50 m.p.h. in the mountains, traffic normally moves much faster. [*Richard A. Beal*]

Trucks must keep below a 35 m.p.h. speed on the downhill sections of Highway 17. [*Richard A. Beal*]

to just north of the Cats. Southbound its from the summit to Granite Creek Road. The truck limit was lowered in 1961, again in 1963 on the Santa Clara side, and in 1973 on the Santa Cruz side.

There is no minimum speed limit and heavily loaded trucks and cars pulling trailers are often found traveling at 25 m.p.h. or less.

These differences in traveling speed on a two lane road often contributes to accidents because cars traveling near the speed limit change lanes rapidly and frequently to avoid slower drivers.

Mileage Markers

On many California roads, including Highway 17, Caltrans installs mileage marker signs to help maintenance workers and emergency personnel locate specific positions on the highway. You may have noticed small white rectangular signs alongside the road painted white . At the top of the sign in black paint is the number of the highway, an abbreviation for the county and the exact mileage from a given spot (normally the county line). SCL stands for Santa Clara County and SCR is the abbreviation for Santa Cruz County.

Going southbound from the 280 intersection, the signs are labeled SCL - and then the mileage to the county line (summit) . For example 10.45 means you are almost ten and a half miles from the summit. Southbound at the summit where

This state roadside marker sign indicates that it is on Highway 17, in Santa Cruz County, and that it is 5.91 miles to the Santa Clara County line at the summit. [Richard A. Beal]

139

you cross into Santa Cruz County, the signs begin with 12.55 and continue down to the Highway 1 intersection.

Northbound from Highway 1, the signs start with 0.00 and proceed to increase until the summit where you cross into Santa Clara County. There they again begin at zero. From Santa Cruz to the summit is 12.55 miles, from the summit to 280 is 13.82 miles.

Lane Numbering

Lanes are numbered from the inside fast lane to the slow outside lane. Thus the lane closest to the center divider is 1, and the slow driver lane next to it is lane 2, etc. Use the correct lane number when reporting accidents or incidents to assist those responding to emergency services.

On Ramp Metering Lights

In the 1970s traffic lights were installed at northbound on-ramps for Camden, Lark and Hamilton. The intent is to restrict the number of cars that can enter Highway 17, thus keeping the main traffic flow going. The lights only work during the week from 6-9 a.m. Los Gatos and Campbell residents hate the lights because they cause long delays and back traffic up on city streets.

APPENDIX 2

Maintenance

Introduction

Maintenance and construction on Highway 17 is the responsibility of the State of California Department of Transportation (Caltrans). Specific responsibility lies with District IV of Caltrans, 3333 California Street, San Francisco, (415) 934-4444. Although some contractors are used, the bulk of the work is performed by 137 state employees who work out of the San Jose District IV office, 500 Queens Lane, San Jose (408) 436-0930. Those individuals have responsibility for state roads in all of Santa Clara County and some other adjacent areas such as Highways 1 and 17 in Santa Cruz County. About half of the people are performing landscape maintenance related activities - the rest are concerned with the road surface, signs, drainage, barriers, etc.

Landscaping along Highway 17 is minimal compared to some sections of road in California, and lack of funding is blamed. For the San Jose office, each landscape maintenance person is responsible to maintain an average of 30 acres of landscaping!

Incidentally, if you cause more than $100 of damage to state property (e.g., damage a barrier or sign), Caltrans will eventually bill you for repair costs.

There is increasing emphasis on preventive work - for example Caltrans pays special attention to sealing cracks on the highway. Keeping the road surface sealed means that water can't work its way under the roadbed and cause further erosion.

In the winter time, emphasis is placed on keeping the drainage system working properly. For example, the concrete barriers have small openings that allows water to drain through - but these become clogged with dirt and debris.

Whenever freezing temperatures are projected, Caltrans has a special team in place at the summit to use sand or snowplows to keep the road clear. Sand is stored at several of the larger turnouts and is spread on the road to help improve traction during icy conditions. Salt is not used on Highway 17.

Caltrans is also responsible for picking up dead animals and disposing of them safely.

Snow on Highway 17 in January 1949 caused huge traffic jams. This picture was taken near Pasatiempo Drive. Notice the old state highway sign in the lower right corner. *[UCSC Special Collections]*

A two mile section of Highway 17 near Los Gatos has been adopted by Nancy Walker and her friends who will keep the area free of litter. This volunteer program has been very successful, according to Caltrans. *[Richard A. Beal]*

Litter and debris is a large problem along California roads - but it is getting better. Experts believe that today we are less likely to litter because of increased environmental awareness. When people do litter (and it's a $1,000 fine if you are caught!) Caltrans must pick it up. Crews regularly perform this task, along with citizens who elect to help work off civic and traffic violations instead of paying a fine. Inmates also sometimes help.

There is a new "Adopt A Freeway" program which has had great success in the last 3 years. Community groups adopt a 2 mile section of road and agree to have trash pickup events at least 4 times a year. If you are interested in learning more about this contact Ed Costa, Caltrans, at (408) 436-0930.

Working on the highways is a dangerous job. More Caltrans workers have been killed "in the line of duty" than CHP officers. Although they carefully put warning cones to close lanes and normally have flashing lights, poor drivers still manage to hit people.

Caltrans Levels of Service

In the 1960s Caltrans adopted a common series of terms to help define the traffic levels on various highways. Although these are open to some interpretation, you will find them used in various official transportation documents. As you might guess, Highway 17 is normally described as having a D Service level, going to F by 1995. Some commuters would argue that it is already at F.

Level of Service A

Free flow, low volume, high operating speeds, high maneuverability, little or no delays at intersections. (volumes up to 60% of capacity)

Level of Service B

Stable flow, moderate volume, speed somewhat restricted by traffic conditions, high maneuverability, short traffic delays at intersections. (volumes from 61% to 70% of capacity)

Level of Service C

Stable flow, high volume, speed and maneuverability determined by traffic conditions, average traffic delays at intersections. (volumes from 71% to 80% of capacity)

Level of Service D

Unstable flow, high volumes, tolerable but fluctuating operating speed and maneuverability, long traffic delays at intersections. (volumes from 81% to 90% to capacity)

Level of Service E

Unstable flow, high volumes approaching road capacity, limited speed (less than 30 m.p.h.), intermittent vehicle queuing, very long traffic delays at intersections. (volumes from 91% to 100% of capacity)

Level of Service F

This category is for locations where a traffic problem downstream causes a backup of traffic. The result is forced flow, volumes lower than capacity due to very low speeds, heavy queuing of vehicles, frequent stoppages.

Reporting Road Problems

If you notice maintenance-related problems along the highway of an emergency nature, call 911 or use one of the emergency call boxes. An emergency is anything that is blocking the road or endangering human life. If it is not an emergency, call Caltrans at (408) 436-0930.

1989 Earthquake Damage

Earthquakes have damaged the roads over the Santa Cruz Mountains from the earliest days. The latest was a massive 7.1 earthquake in 1989 that was centered less than 10 miles from parts of Highway 17. Although the road was totally closed only a few days, the traffic backups continued for 32 days with up to 3 hour commutes each way(!) not unusual. Alternate routes were horribly clogged. Caltrans used its emergency funds to hire contractors to help out with the repairs and performed near miracles working around the clock to get the highway open again.

The 1989 repairs cost $5 million and involved moving 27,000 truck loads of dirt that had fallen onto the road in landslides. The concrete center barrier had hundreds of breaks in it and there were large cracks in the pavement. Near Sims Road portions of the highway were raised several feet by the earth movement.

APPENDIX 3

Safety Equipment

The "Jersey" Concrete Barrier

The greatest single contribution to surviving a trip over Highway 17 has been the concrete divider between the two opposing lanes of traffic. The so called "Jersey" barrier runs along most of the length of Highway 17 with the first sections put in place in 1984. The CHP credits it with saving dozens of lives a year. Yes, every mark on the barrier is a spot where a car has hit or brushed the wall - most likely with their bumper. Each 20 foot section of the concrete wall weighs 6 tons and each section is tied to the next with steel reinforcing rods. Some of the newer sections of the wall are actually continuously poured concrete. The peculiar angle of the wall is designed to push the car away at a shallow angle in the same direction it is traveling. At the end of a Jersey wall section, plastic or metal barrels are placed to protect cars. The barrels are filled with sand, or occasionally water, to help absorb shock if they are hit.

During the 1989 earthquake in the Santa Cruz Mountains, the Jersey wall had 350 breaks or significant cracks in the Highway 17 section alone!

Stripes and Dots

There are two types of lane markers on Highway 17 - painted lines and Botts disks - those round raised ceramic dots. The Botts disks are fastened to the road

147

using a special epoxy glue. Transportation system designers have learned that the dots are easier to see in poor weather and because you can feel them as your tires move over them, they help people keep in their proper lanes.

Daylight Safety Program

The daylight safety program is a series of signs near Granite Creek Road and the Lexington Reservoir that ask motorists to turn on their headlights as they transit the mountain portion of Highway 17.

The idea is that this makes it easier to see other vehicles. The CHP strongly endorses the program and says that it definitely works. You should use only your headlights, however, as it is illegal in California to drive with your parking lights on.

Don't forget to turn your lights off when you are out of the mountains!

Accident Warning Sign

Only one permanent information sign exists on the highway. Just north of the summit, after the Summit Road exit, Caltrans installed a warning sign that is manually activated by the CHP. Normally closed, it can be opened to reveal the words "WRECK AHEAD" and flashing yellow lights are turned on. The CHP also uses the sign as a warning when the road is slippery or there is a slowdown

The daylight safety program began before the center divider was in place. Many motorists still follow it and the CHP claims that it definitely decreases accidents by increasing the visibility of your car to other drivers. *[Richard A. Beal]*

due to construction. Remember that the next one-half mile of downhill road has the highest number of accidents, so your risk is high. Be careful when you see the sign!

The Scotts Valley CB REACT club lobbied for a permanent sign in Scotts Valley, northbound near the Granite Creek exit. They wanted to warn motorists of problems or delays so that they could use alternate routes. Caltrans opposed the idea saying a sign was too expensive.

911

The 911 emergency phone number is used anywhere along Highway 17. Depending upon where you are along the highway and which type of phone you are using, you will be connected with different 911 centers. In all cases, however, there are specially trained personnel who can help you and dispatch the proper emergency personnel. The 911 system should only be used for true emergencies.

Public Telephones

There are several public telephones along Highway 17. The Detailed Listing section at the end of this book gives you the locations depending upon your direction of travel.

Emergency Call Boxes

The locations of emergency call boxes are listed in the Detailed Listing section of this book. There are two listings, one for each direction of travel.

In early 1991 the Service Authority for Freeway Emergencies (SAFE) motorist aid call box program was put into effect. These motorist aid call boxes are spaced between one-half and 1 mile apart the entire length of Highway 17. You can recognize the call boxes by their large blue and white signs (the call boxes themselves are yellow and mounted on the pole at chest height).

Inside the box is a telephone that directly connects you to the CHP (California Highway Patrol) police dispatcher in either Vallejo or Monterey, depending upon your location. Funding for the call boxes come from a $1.00 per year fee on your car registration.

Most of the call boxes use solar power to charge an internal battery (like a motorcycle battery). Cellular telephone technology is used to reach the CHP dispatcher. Cellular One operates and maintains the system. THE CALL BOX IS ONLY FOR EMERGENCY CALLS. It is not a regular telephone. Use the call boxes only if your vehicle breaks down or you wish to report a hazard or an

Emergency call boxes are located approximately every one mile along Highway 17. Each box uses solar panels to recharge batteries, and uses cellular telephone technology to reach emergency operators. *[Richard A. Beal]*

accident. Opening the box will set off an alarm in the dispatchers office to detect theft attempts or vandalism and police will quickly response to the location. NEVER CROSS THE FREEWAY TO REACH A CALL BOX!

To use the phone:

1. Open the box

2. Lift the telephone from the cradle and wait for the operator

3. Stay on the line, even if the call is not answered immediately

4. Speak clearly to the operator

5. Remain calm and follow the operator's instructions.

6. Replace the receiver when finished and close the box.

Although the dispatchers can automatically tell which box you are calling from, you can also identify your location from the sign above the call box. The first two letters identify the county (SC - Santa Clara, SZ - Santa Cruz), the next two or three numbers the highway number (17) and the next two or three numbers are a unique number for that call box.

Radio Frequencies

CB channel 17 is the primary place to hear information about Highway 17. Truckers use this channel almost exclusively and the area volunteer REACT teams monitor it along with the channel 9 emergency channel.

The CHP uses 42.24 MHz. to communicate between the dispatcher and patrol cars. They still use VHF radio frequencies and have no immediate plans to move to the higher frequency 800 MHz bands. The VHF frequencies work better in the canyons of Highway 17.

Hams commonly use repeaters:	146.115+	(WB6ADZ)
	224.720-	(KA6S)
	440.550+	(KB6GBX)
	440.850+	(N6IYA)

Cellular Phones

Cellular phones are increasingly popular and can be used along Highway 17. Reception in the mountains is often erratic so expect some problems, but in general they work just fine. Many commuters have installed them in their cars so they can use their time more productively, as well as in case of emergencies.

GTE was the choice of most local dealers contacted, they claim it gives the most reliable coverage over the Santa Cruz Mountains.

Radio Station Information

Many radio stations have "traffic reporters" these days. They get their information from listening to police scanners, cellular telephone reports, airplane spotters, etc. The CHP will tell the press if they believe any incident will cause traffic backups of more then 20 minutes. The stations vary in the quality of their information, but generally the all-news AM stations have the best coverage: KGO (810) and KCBS (740).

To use the emergency phone, open the cover and simply pick up the phone. Directions are printed inside in four languages. *[Richard A. Beal]*

Holy City around 1930. The studio and radio tower for Riker's AM radio station KFQU appear at the top. *[William A. Wulf Collection]*

APPENDIX 4

Problems while on the Highway

Introduction

The greatest danger on the highway is the possibility of being hit by other vehicles. This applies to those directly involved in accidents or with disabled vehicles - and those who stop to provide aid. Think safety!

You are required by law to warn follow-on drivers if there is a hazard. If you see an accident, you are required by law to stop in a safe location, render aid and provide your identification to the people involved. You are protected by the Good Samaritan Law (see following).

The first thing to do is move the vehicle to a safe location - if you possibly can. This means a spot where the car is completely off the road and can be seen by cars at least 500 feet back. The only exception is never to move the car if someone is seriously injured.

If you have a problem, the CHP advises that in daytime you lift the hood and then put on the emergency hazard blinkers or your turn signal on if you don't have hazard lights. Then get in the car, roll up the windows, lock the doors and stay inside the car until the police arrive. Do not open your window or door to talk to people. If people do stop, ask them to call 911 and report the problem. Do not go with anyone, stay with the car.

If you are reporting the problem and a woman is involved, inform the dispatcher. The CHP gives higher priority to accident calls with women motorists (known as an 11-26 response).

At night, women should follow the same guidelines: hood up, hazard lights, stay inside with the door locked. But the CHP also recommends that you "become invisible" by lying down on the seat.

Recommended Emergency Equipment

For Your Car Trunk

- 5 pound ABC fire extinguisher
- 6-12 flares
- First aid kit
- Flashlight in the glove compartment (helps if it works.)
- Jumper cable for battery starting
- Can of instant tire inflator (for flat tire)
- One gallon of distilled water in a plastic container (can be used in radiator, to cover burns, fight fires)
- Old blanket
- Pair of old tennis shoes (especially for women with high heels)
- This book!

Getting Help

You can use cellular phones to call 911, the emergency call boxes located every half mile or so or use regular pay phones. Try to be precise when saying where the problem is located - emergency response personnel say this is one of their biggest problems. Give the direction of travel, use the closest mileage marker sign, use the maps in this book. Be precise, it's important!

Danger of Being Hit

CHP and fire rescue officials all tell stories about drivers who go right through a line of flares and flashing lights - and hit stopped cars or people at the vehicle site. Remember, your greatest danger is being hit. Don't assume that drivers can see you.

Move To A Safe Place

If no one is seriously injured, move the car to a safe place on the side of the road. The CHP highly recommends this and says that not moving the cars until the police can see the accident scene is an old wives tale. Move your car to safety!

Make sure oncoming drivers can see your car in time. Park it completely off the road if at all possible. If not, you need to put the hood up, turn on your hazard lights or a turn signal if you don't have hazard lights. Get away from your car to the safest possible place. For some reason, drunk drivers are especially attracted to bright lights and may hit your car.

Flares

NEVER LIGHT A FLARE IF YOU SMELL GASOLINE!

Safety flares are highly recommended to inform on-coming drivers of a problem. If your car is on the road or even close to the road, put out flares.

Flares can be purchased at any automotive supply store or most supermarkets.

They burn according to their length - the longer they are, the longer they burn. Don't save a few pennies by buying the shortest ones - they may save your life someday. Put 6 flares in the trunk of the car - and then put 6 flares in a towel or plastic bag and put them under the drivers seat. Rear end collisions are common on Highway 17 and you may not be able to open your trunk.

How To Safely Light Flares

Flares are dangerous and must be treated with respect. They burn at 2,000 degrees and can burn you badly.

Never light a flare if you smell gasoline or oil! And don't put the flare near brush that might catch fire. Remember that gasoline is a liquid and even though the vehicle(s) may be several hundred feet away, there may be a gasoline pool on the highway. Look carefully and smell for gasoline BEFORE you light a flare.

The flare is a long tube with a cap on it. Remove the cap by pulling or twisting. Hold the cap in one hand as far away from you as possible, with the top of the cap towards you. Hold the top of the flare (the part that just became exposed) in the other hand. Turn your head to the side so sparks won't fly in your eye. Holding both hands away from you, strike the flare top against the cap top like you would strike a match. Rubbing the two parts together will normally start the flare. Continue to hold your hands out and slowly stick the cap on the opposite end of the flare. Continuing to hold the flare far away from you, slowly put it down on the ground. Never throw the flare down because it can break or shoot sparks out.

Flares burn from 10-20 minutes depending upon their length. It often will take longer than that before the CHP can respond so you need to monitor them and replace them when necessary.

If the hazard is gone, let the flares burn out naturally. Don't try to put them out.

Wigwam

If you have enough flares, you can use a trick called the wigwam. You remove all the caps, put 2-3 flares at 45 degree angles to one another with the top of one about 3" from the bottom of another flare. As the first flare burns down, it ignites the next flare.

Where To Put Them

The most common error with flares, and other warning devices, is not to put them far enough away. Cars are coming towards you at 55 m.p.h.–that's 80 feet every second! Carefully walk back from the accident scene or disabled vehicle about 500 feet. That's a long way, about 200 paces for the average person. Put the first flare there. Then continue to walk back towards the car, placing flares about every 100 feet until you have reached the accident scene. Your last flare should be here. Put the flares in a straight line, at an angle across the lane, helping to direct drivers into the safe lane.

Triangles & Lights

If you don't have flares, the red reflective triangles, flash lights, etc. can all be used but be extra careful. These normally cannot be seen from a distance and therefore are not much use - but they are better than nothing.

Injuries and First Aid

If you are arrive at an accident scene or someone having a medical difficulty, there are some basic rules to follow.

Get Help

People with medical problems need professional help rapidly. Stop passing cars and ask them to go to the nearest call box and request medical help. Truckers can call on their CB to radio REACT teams. Cars with cellular phones can call 911. It's a good idea to ask several people to call - to make sure at least one does. Give them them an accurate location, use the mileage marker signs if possible.

Don't Move Injured People

If people are injured in an accident, or are experiencing a medical problem, don't move them unless there is a danger of fire. People are often disoriented and under-estimate their injuries. When in doubt, keep people lying down and still. Help is on the way!

Check for Breathing

Check for breathing. Use CPR when appropriate if you know how. Remember the alternative to doing nothing is probably rapid death for the person. CPR has saved many lives and the training is highly recommended. Contact any fire department or hospital for information on classes in your area.

Stop Bleeding

Control bleeding by applying direct pressure to the bleeding area. Use a clean cloth, or your shirt or anything you can find. It's very important to stop the bleeding. Tourniquets can be effective if you know how to use them - but direct pressure is surer, faster and safer.

Keep Warm And Reassure Them

Shock is very common in accidents as the body throws its resources into fighting the accident effects. Body temperature is reduced. Keep people warm by covering them with a blanket or jacket - or anything that will trap their body heat. Keep reassuring them that help is on the way.

Burns

If the victim is burned, be extremely careful. If you have distilled water, gently pour the water on the burns. Don't remove clothing. It's OK to pour water directly on the clothes over the burn area. This will help reduce the pain and disinfect the skin.

Fires

Most vehicle fires will be in the engine compartment. Firemen interviewed said it is rare for a vehicle to explode, but play it safe and get away from the car. Cars can be replaced - you can't. If the car doesn't catch fire at the time of impact, it probably won't. The reality is that if your car is already on fire, the fire department probably can't respond in time to save it so focus on getting people to a safe spot at least 100 feet from the car. And avoid getting close to the front or rear of the car. The new shock absorbing bumpers have shock absorbers that can, and often do, pop when heated - suddenly expelling the bumper. Also the large amount of plastics in today's automobiles can cause toxic smoke.

Fire Extinguishers

Any fire extinguisher will be useful but the firemen who regularly respond on Highway 17 all recommend a basic dry powder ABC extinguisher. Any size from 5 to 20 pounds is appropriate.

Remember that fire extinguishers are pressurized to expel the fire retardant powder - and that they can lose their pressurization. Check the attached gauge each time you change the oil in your car and make sure it will still work!

How To Use A Fire Extinguisher

The most important thing is to read the instructions ahead of time and know how to use it. In most cases you will pull out a safety pin and press down on a lever. Aim the extinguisher at the base of the flame. If the fire is in the engine compartment, open the hood (be very careful as it may be hot or the fire may flash). Just pulling the hood release latch may open it enough to fight the fire. If you smell gasoline, get away from the car.

Chemical Spills

A new hazard on Highway 17 is chemical spills. In 1990 a 4,000 gallon gasoline spill in Scotts Valley closed the road for many hours. If you come upon an accident scene and notice any type of powder or liquid spill - get everyone to backoff. Not only are there gasoline trucks on the highway, but many industrial acids are carried for the tannery plants, Lockheed, etc. Even a grocery chain truck can have bleach and other common materials which can react together in dangerous ways. When in doubt, back away. All the fire departments have hazardous material response teams and only they can determine what is safe.

Good Samaritan Law

In California, there is a law commonly called the Good Samaritan Law. If you stop at an accident scene, use recognized first aid and do not exceed your abilities, California law will protect you from later law suits claiming your actions might have caused further injuries. Common sense prevails.

Accident Reports

You must fill out a Department of Motor Vehicle accident report if anyone is injured or if the total related damage to vehicles and/or property is over $500. Go to any DMV office and request form SR-1. If a CHP officer was at the accident scene, they will normally prepare their own accident report. You can get a copy at their office after a week has passed.

Vehicle Disabled

Again, the first thing to do is to get your vehicle to a safe place and put out warning flares. You can use 911 or the emergency call boxes to call for help. Most tow trucks require cash or a credit card.

Several emergency road side are services offered through the California State Automobile Association, Mobil Oil, GTE Mobilnet, etc. Costs are typically $30-50 per year for emergency road service. Just one call will more than pay for itself.

Don't allow your car to be towed to a repair shop you don't know. Don't authorize any repairs by signing a towing release unless you have decided to have your car repaired by the shop where your car is being towed. Don't feel pressured by the driver. You can decide about repairs later when you are calm.

Deer are seen on Highway 17 each year. *[Richard A. Beal]*

NEXT
4 MILES

Some tow truck drivers carry spare gasoline, water, hoses, etc. but not all have repair parts. Remember that the most common call on Highway 17 is for overheating.

If you get a flat tire, be especially careful if you change it with the spare. You don't want your rear end sticking out into a traffic lane with 80,000 lb. trucks going by at 55 m.p.h..

Injured Animals

There is a considerable deer population living in the Santa Cruz hills and unfortunately it is not uncommon for one to try crossing the road. Dogs also occasionally wander onto the road.

Because of the speeds and lack of visibility, there is little warning to drivers. Hitting a large dog or deer can cause a serious accident and people have been killed. In 1990 drivers reported accidents involving 10 deer and 4 dogs on

Highway 17 - but commuters know there are many more animals killed each year that are not reported.

There is not much you can do to avoid such accidents. Remember that the greatest hazard to you is going over the side of the road. It is better to hit the animal than go off the road. The CHP puts a device called a "deer warning cycle" on their vehicles. It is a small metal device which mounts on the front bumper of the car. When wind goes through it, it produces a high frequency pitch which is supposed to irritate animals and cause them to move away. You can buy them at Orchard Supply Hardware stores. CHP officers are convinced it really does help.

If you injure an animal, first get to a safe place. Use the emergency call box or a phone and call 911. They will dispatch the SPCA to provide first aid. Also call if you kill an animal, so it can be safely disposed of.

Chains

The CHP can require that you put chains on your car when the road is icy. Although not a common occurrence, this has occurred several times in the past 10 years. Granite Creek exit area (northbound) and Black Road (southbound) are the points at which the CHP normally halts traffic for chain requirements. As you might expect, this causes *severe* traffic backups and might be a good excuse to stay home and sip hot chocolate.

Hazardous Drivers

The guy in the big truck is chasing the red Porsche who cut him off. The drunk driver can barely keep from hitting the barrier. What should you do?

The first thing is to back off and get some distance between you and the other driver. If you have a cellular phone or can get to a call box, call 911 and report the problem. Maybe the police can stop the car.

Unfortunately, just getting their license number doesn't do much good. The police can't issue citations unless they personally see the violation. If you want to pursue the matter through a "citizen's arrest" you can do so, but most people are reluctant to get involved.

If you do have their license number, you can call the administrative CHP phone numbers and ask that a warning letter be sent. The CHP will look up the vehicle registration information and send the registered owner a letter saying "your vehicle was seen Thursday at 5 pm driving in a reckless manner. Please be aware that the law requires, etc....". At least you'll feel better.

Traffic Tickets

Despite our best intentions, sometimes it does happen. You look in the mirror, see that flashing red light and feel your pulse rate double. Be very careful where you pull off - make sure there is enough room for both cars to safely get completely off the road. Most CHP vehicles have loud speaker systems and the officer will normally give you directions on where they wish you to stop. Turn off your radio, roll down a window and listen carefully to what they are saying.

If you are stopped, stay in your car unless the officer gives you other directions. You will be asked to provide your drivers license and vehicle registration certificate. Officers normally check your drivers license and registration with the DMV computer databases to see if they are current and valid.

If you have any outstanding warrants for minor items or unpaid tickets, they normally will follow you to the nearest courthouse so it can be cleared. Drunk drivers, unlicensed drivers and those with suspended licenses go directly to jail and their car is impounded.

Several CHP officers stated that 1 in 5 drivers they stop do not have valid drivers licenses.

Traffic tickets received in Santa Cruz County are handled by the Santa Cruz County Municipal Court Room 060 at 701 Ocean St. 8 am-4 pm, Santa Cruz (408) 425-2324. Traffic tickets received in Santa Clara County are handled by the Santa Clara County Municipal Court 8:30 am - 4 pm, 935 Ruff Drive, San Jose (408) 299-2233. There is also an office in Los Gatos at 14204 Capri Drive (408) 378-3408 and Sunnyvale at 605 W El Camino Real (408) 739-1676.

Impounded Cars

To retrieve an impounded car, you need to clear up the initial problem, take proof to any CHP office and obtain a release form. Take all the paper work and go with a licensed driver to the tow truck company that is storing your car. You must show them the CHP release form and pay the storage fees in order to get your car back.

APPENDIX 5

Alternate Routes

Introduction

The following map shows the significant alternate roads that can be used off Highway 17. Be aware that most of these are two lane, very winding narrow roads which may not save you any time. But they certainly are scenic!

Map of
Alternate Routes

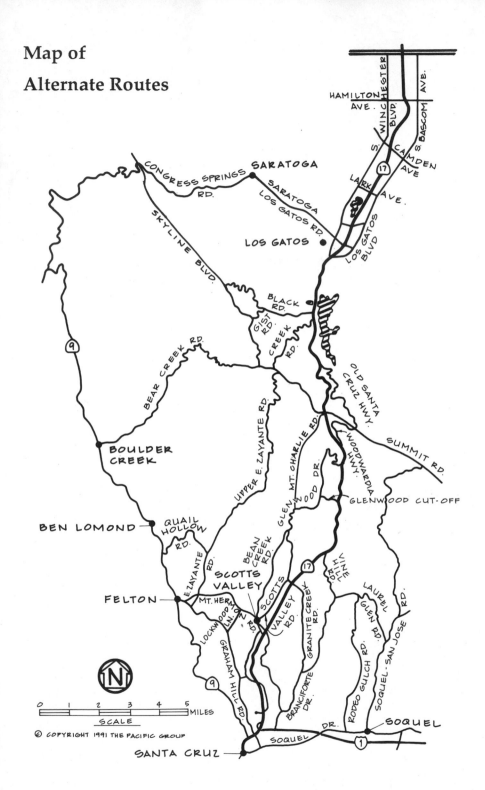

APPENDIX 6

Services

Introduction

There are many businesses along or near Highway 17 and the following list is certainly not comprehensive. It does, however, list many of those services closest to Highway 17. Owners of adjacent businesses are welcome to send information for inclusion in the next edition of this book.

Places to Stay

Pruneyard Inn and Campbell Inn

Both the Pruneyard and Campbell Inns are owned by the same people. Manager is Dermot Connolly. The Pruneyard Inn is located at 1995 S. Bascom Ave., Campbell 95008, phone (408) 559-4300. The Campbell Inn is at 675 E. Campbell Blvd., Campbell 95008, phone (408) 374-4300. There is also a toll free reservation number for both Inns, (800) 582-4300. The Campbell Inn has 95 rooms with prices from $89-$175. The Pruneyard Inn has 116 rooms with prices from $89-$250. Both have upgraded rooms with fireplaces and jacuzzi tubs. Both also have optional VCRs and a movie rental library. All guests can use the swimming pools, public spas and tennis courts. A free continental breakfast is included as is a free shuttle for local transportation. Guests can use free mountain bicycles and the Pruneyard Inn is next to the 13 mile jogging trail along Los Gatos creek.

Both Inns also have banquet and meeting rooms available. Check out is noon, check in is 2 p.m., all major credit cards are accepted. The Pruneyard Inn is adjacent to the Pruneyard shopping center which includes 13 restaurants. Both Inns have been rated "4 stars" by the AAA.

Los Gatos Lodge

This famous hotel was opened in 1958. Located at 50 Saratoga Avenue in Los Gatos, phone (408) 354-3300, the 125 room establishment is located on 9 acres of woods, lawns and beautiful gardens. Room prices run from $64-129. They have a pool, spa, shuttle to the San Jose Airport and non-smoking rooms. Major credit cards are accepted. In addition there are 5 meeting or banquet rooms and many patio areas. The Lodge is popular for weddings and receptions. There is a restaurant and cocktail lounge on the property. The Manager is Michael Clausen.

The Toll House Hotel

Located at 140 South Santa Cruz Avenue, downtown Los Gatos, phone (408) 395-7070. There are 97 rooms with prices from $89-175. A few suites are available. There is room service, banquet and meeting rooms and a complimentary continental breakfast Mon-Fri. Guests can use the nearby Los Gatos Athletic Club without charge and historic downtown Los Gatos is within walking distance. They also have a free shuttle bus to the San Jose Airport. Inside the hotel is a restaurant and lounge. Visa, Mastercard, American Express and Diners Club cards are accepted. General Manager is Carole Lee.

In the 1920s Alma was a favorite stopping point for thirsty motorists out for a Sunday drive. [William A. Wulf Collection]

The Cats Restaurant as it looks today. Notice the two small cat replicas of the larger cats at Poet's Canyon standing guard at the front door. *[Richard A. Beal]*

Scotts Valley Best Western

Located at the north end of Scotts Valley, to the west of Highway 17, at 6020 Scotts Valley Drive next door to Denny's Restaurant. Take the Granite Creek Road/Glenwood Drive exit, go west, then north on Scotts Valley Drive. Phone is (408) 438-6666 or for reservations (800) 528-1234. They have 58 rooms, 16 non-smoking, handicapped rooms and are open 24 hours. AAA approved. Visa, Dinners Club, Discover, Mastercard and American Express cards are accepted - no personal checks. Rates are around $60 per night with military, senior and corporate rates available. Guests can enjoy a swimming pool, hot tub, sauna, cable TV, guest self serve laundry, conference room and free coffee served in the morning. Check in in 2 p.m., check out at 11 a.m. Owners are Bob and Diana Hogan, Marlyn and Bonnie Bergman.

The Inn at Pasatiempo

Take the Pasatiempo Drive exit and the Inn is located on the west side of the highway at 555 Highway 17, Santa Cruz, Phone (408) 423-5000 or (800) 834-2546 within California. They are 2 miles from the Santa Cruz Boardwalk and have 51 rooms plus 3 suites at prices from $75-175 per night. American Express, Visa, Mastercard, personal checks with ID accepted. Extras include jacuzzi tubs in the suites, TV, VCR, cable TV, swimming pool, complimentary European breakfast from 7-10 a.m. Directly behind the Inn is the Pasatiempo Golf Course, rated one of the top 10 public courses in the U.S. Check in is 3 p.m., checkout is noon. There

Inspiration Point on the old road, before Highway 17 construction widened the area and eliminated the snack stand. *[William A. Wulf Collection]*

is a cocktail lounge and restaurant at the same location. The Inn is owned by the Property Service Corporation.

Restaurants & Cocktail Lounges

Campbell has 13 restaurants alone in the Pruneyard shopping center alone, and many more adjacent on Bascom Avenue.

Los Gatos has many places, especially on Santa Cruz Avenue.

Garden Court Restaurant and Cocktail Lounge

Located inside the Los Gatos Lodge at 50 Saratoga Avenue, Los Gatos, phone (408) 354-3300. This popular restaurant features a wide variety of high quality dishes. The cocktail lounge has live entertainment Thursday, Friday and Saturday plus happy hours from 5-8 Monday through Friday. And on Fridays there is a special BBQ that is excellent.

Le Restaurant and Lounge

Located inside The Toll House Hotel at 140 South Santa Cruz Avenue, downtown Los Gatos, phone (408) 395-7070. Manager is Jim Zimmerman. The dining room seats 64 and serves breakfast from 6 - 11 a.m., lunch from 11 a.m. to 2 p.m., dinner from 5 to 10 p.m. Friday and Saturday, 9 p.m. the rest of the week. They also have a popular Sunday brunch. The lunch menu includes sandwiches, fish,

168

fettucine, chicken and salads from $5-12; the dinner menu is fancier including linguini and clams, duck, chicken, prime rib, steaks and lobster from $13-20. The adjacent lounge often has entertainment on weekends and has a large screen TV.

The Cats Restaurant and Tavern

Just south of Los Gatos is a well known landmark on the highway. Located at 17533 Highway 17, Los Gatos, Phone (408) 354-4020. Their dinner menu features steaks, BBQ beef, chicken with honey glaze and pork ribs at prices from $8 to $15. All meat is cooked over an oakwood fire. Children's portions are available. The

Russell Aero Foto, S. F., Calif.

The First Airplane View of the Redwood Estates

The winding, paved State Highway may be discerned in the foreground—and above it the Redwood Estates in all their billowy loveliness.

To reach this wonderland in the Santa Cruz Mountains, just drive along the State Highway towards Santa Cruz—and, six miles out of Los Gatos, turn in where the Dutch Mill marks the entrance of the Estates.

Or just drop a postcard—or 'phone in—to one of our offices and we'll take you in one of our bonded cars, with a careful, courteous driver—without cost, without obligating you in any way.

This is a trip worth taking!

REDWOOD ESTATES CO.
Harry W. Grassle and Associates
owners and developers of the
REDWOOD ESTATES
in the Santa Cruz Mountains
OUR FIVE OFFICES
1182 Market Street, San Francisco . Hemlock 7303
410 Fifteenth Street, Oakland . . Glencourt 6823
Burrell Building, San Jose . . . San Jose 8287
48 Santa Cruz Ave., Los Gatos . . Los Gatos 439
556 Emerson Street, Palo Alto . . . Palo Alto 388

Think of all the comforts and conveniences of home in the midst of this mountain paradise—and you have the true view of the Redwood Estates!

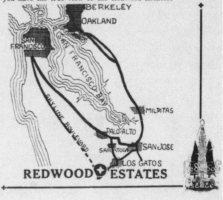

This early photograph of Redwood Estates was published in June, 1927. Notice the phone numbers in the lower left hand corner. *[Cabinland Magazine]*

building has an eclectic style with a country & western flavor and seats 55 in the restaurant, 25 in the lounge. The adjoining lounge has entertainment every night except Tuesday. Hours are 4:30 p.m. to 1:30 a.m., dinner is served from 5-11 p.m. and they are closed on Mondays. Mastercard and Visa cards are accepted. The owner is Evea Ogilvie.

There are two restaurants just south of Patchen Pass:

Mountain Top Restaurant (Formerly Cloud 9)

This is on the east side of the road and was opened in February 1991 by owner Tony Hwang. Hours are 5 a.m. to 10 p.m. daily. Their address is 22990 Freeway 17, Los Gatos, CA 95030 (408) 353-5443. They have take out service, salad bar and an outside BBQ during good weather. The menu consists of steaks, hamburgers, seafood, pasta, pancakes, omelettes in the $4-11 range. Beer and wine are available. The restaurant has 80 seats, a non-smoking area and lots of parking. Restrooms are for customers only. There is also a public phone in the parking lot.

Summit Garden Inn

Located on the west side of the highway, this restaurant is owned by Jimmy Pang. Before 1970 it was known as Hoeffler's. Their address is 23123 Los Gatos-Santa Cruz Way, Los Gatos, CA 95030 (408) 353-2524. They have takeout service, serve breakfast, lunch and dinner with eggs benedict, omelets, pancakes, waffles, burgers, sandwiches, seafood, pasta and Chinese food. Prices run from $4-11. Beer and wine are also available. The Summit Garden Inn also has an outdoor patio area that is used in good weather. There are public pay phones in the parking lot and one inside. Restrooms are inside.

In Scotts Valley there are several places to eat along Scotts Valley Drive and Mount Hermon Road including:

Denny's Restaurant

Located at the Granite Creek Road exit, Denny's is immediately north on the west side of the highway at 6014 Scotts Valley Drive, Scotts Valley (408) 438-2023. They are open 24 hours a day, have breakfast, lunch and dinner menus and can seat 110 people. A non-smoking section is available and it is wheel chair accessible. Food to go is available. The menu is in the $4-10 range with omelettes, ham, seafood, steaks, hamburgers and the other usual fare. Shoes and shirts are required. They have a rest room and public telephone inside. VISA, Mastercard, American Express are accepted. The restaurant is owned by Ken Bergman.

Cafe Carlos

Behind Dennys at the Granite Creek Road exit - immediately north on the west side of the highway. Hours are 11:30 a.m. - 8:30 p.m. M-Thur; 11:30 a.m. - 9:30 p.m. Friday and Saturday; 1 p.m. - 8:30 p.m. Sundays. This is a nice little restaurant with seating for 58 inside and 40 outside on their patio. They feature Mexican food with all the usual items, plus very good mesquite grilled specials. Try the mesquite-grilled red snapper ($10.95)) or shrimp salad ($6.95). Beer and wine are also available along with low alcohol margueritas. Meals run from $5 to 12 and they only accept Mastercard and Visa credit cards. A public phone is outside and a rest room inside. Shoes and shirt are required. The owner is Carl Taylor. Cafe Carlos is located at 6016 Scotts Valley Dr., Scotts Valley (408) 439-8448.

Kentucky Fried Chicken

About 1 mile west of Highway 17 at 95 Mt. Hermon Road, phone (408) 438-2633. Hours are 10:30 a.m. - 10:00 p.m. daily. They have the usual chicken, bread, gravy, cole slaw, soft drinks at prices from $3 to $19. The restaurant seats 30 people and there is also a drive through. Restrooms are inside and a pay phone is nearby. No checks, no credit cards. The owner is Nick Juliano.

Peachwood Bar

Take the Pasatiempo Drive exit, the bar is located on the west side of the highway at 555 Highway 17, Santa Cruz, inside the Pasatiempo Inn. Phone (408) 426-6333. The bar is open from 11:30 a.m. to 1:00 a.m., no cover charge, live entertainment with a small dance floor on Friday and Saturday nights. They also have a happy hour with snacks and reduced drink prices from 4-7 p.m. Mon-Fri. Wines by the glass are available. The Bar is owned by the Property Service Corporation.

Peachwood Restaurant

Take the Pasatiempo Inn exit, go west and the restaurant is located at 555 Highway 17, Santa Cruz, inside the Pasatiempo Inn. Phone (408) 426-6333. The restaurant can seat 112 inside, with an additional 77 outside on the decks when weather permits. Lunch is served from 11:30 a.m. to 2:00 p.m. (M-F), dinner from 4:30 p.m. to 10 p.m. (9 p.m. Sunday), Sunday brunch from 10 a.m. to 2:30 p.m. Visa, Mastercard, American Express, Carte Blanche and Diners Club cards are accepted. They have a children's menu available and an early bird special menu good until 6:30 p.m. An excellent menu is available featuring California cuisine dishes like fresh salmon, scallops, chicken, lamb chops, pasta primavera, london broil and prime rib. Soups, salads and starters are equally varied. Prices for lunch run from $4-$9 and dinner entries from $7-16. Byron Gravelle is Chef and the restaurant is owned by the Property Service Corporation.

Gas & Diesel Stations

Campbell

There are no gas stations at Lark (there is a Chevron car wash with gas pumps) although at this writing a new British Petroleum station was going in at the corner of Lark and S. Bascom Avenue.

At the Camden exit, to the east of Highway 17:

Beacon located at 1370 Camden, Campbell, phone (408) 377-1234. This is a 24-hour self service station managed by Cesar Bosano. They have a convenience store and take Visa, Mastercard and Discover cards.

Chevron is at 1402 Camden, Campbell, phone (408) 377-2192. This station was being re-built at this writing and no additional information was available.

Los Gatos

To reach all of these stations, take the Highway 9 exit west on Saratoga Avenue about four blocks. There are also stations to the east of Highway 17, south on Bascom, but these are the closest.

Beacon located at 200 Saratoga Avenue, Los Gatos, phone (408) 354-9801. The self-service station is open 24 hours a day. They take Visa, Mastercard, Beacon credit cards and have a small supply of oil and other fluids, maps, cigarettes. Mr. Matia is the Acting Manager.

UnoCal 76 is at 300 Saratoga Avenue, Los Gatos, phone (408) 354-9539. This self service station is open seven days a week from 6 a.m. - 10 p.m. They have diesel fuel besides the usual gasoline and take American Express, Visa, Mastercard, Discover and Unocal credit cards.

Chevron offers both full and self service 24 hours a day at 275 Saratoga Avenue, Los Gatos, phone (408) 354-5910. They also have soft drinks, cigarettes, candy and maps available. A mechanic is on duty from 8 a.m.-5 p.m. Monday through Saturday. Credit cards accepted include Visa, Mastercard, American Express, Chevron, Discover and the new JCB Japanese credit card. Joseph Bielefeld is the Manager.

Shell is open 24 hours a day and offers both full and self serve. The station has diesel fuel and takes Mastercard, Discover, Visa and the Shell Card. The location is 255 Saratoga, Los Gatos, phone (408) 395-3551. A mechanic is on duty Monday through Saturday from 10 a.m. - 5 p.m. They also sell soft drinks, snacks and cigarettes. The Manager is John Glinka.

Scotts Valley

Chevron in front of Dennys at the Granite Creek Road exit - immediately north on the west side of the highway at 6012 Scotts Valley Drive. The phone (408) 439-5140. The station is open 24 hours daily and a mechanic is on duty from 8 a.m. - 5 p.m. There is a convenience store inside, air and water are available outside. Restrooms are inside, there is public pay phone in back of the station. Chevron,

Visa, Mastercard, American Express, Discover and Optima credit cards are accepted. Self and full service. Bill Sockwell is the Owner.

Shell at the Granite Creek Road exit - immediately north on the west side of Highway 17 at Scotts Valley Drive. The actual address is 1 Hacienda Drive, Scotts Valley, phone (408) 438-3344. They are open 24 hours daily with a mechanic on duty from 7 a.m - 5 p.m. (M-F) and 8 a.m. - 4 p.m. Saturday. There is a convenience store inside, rest rooms and a public telephone. The garage is AAA approved and they perform all automotive repairs. Shell, Discover, Mastercard and Visa cards are accepted. Terry Adams is the Owner. Self and full service.

Exxon at the Granite Creek Road exit - immediately north on the west side of Highway 17 at Scotts Valley Drive intersection. The address is 5620 Scotts Valley Drive, phone (408) 438-9916. Discover, Mastercard, Visa and American Express cards are taken. Hours are 6 a.m. - 8 p.m. Monday through Saturday. A mechanic is on duty from 8 a.m. - 5 p.m. Monday through Friday. There is a public phone, rest rooms and they sell propane and kerosene. Self service only. They also have tow truck service available directly at the station. Stan Epple is the Owner.

Shell about one-half mile west of Highway 17 at 90 Mount Hermon Road (corner of Scotts Valley Drive), phone (408) 438-5727. They are open 24 hours a day, have a large convenience store. Restrooms and a public phone are provided. They also have a drive through car wash that is free with a fill up. Full and self service are provided. There is no mechanic on duty. Larry Palmer is the Owner.

Union 76 about one-half mile west of Highway 17 at the corner of Scotts Valley Drive and Mt. Hermon Road. Phone is (408) 438-1617. They are open 5:30 a.m. - 10 p.m. Monday through Friday and 7 a.m. - 10 p.m. weekends. A full service garage is available 8 a.m. - 5 p.m. weekdays and 9 a.m. - 2 p.m. weekends. They take Discover, Mastercard, Visa, American Express and Unocal cards. A public pay phone is nearby and rest rooms are provided. They have both full and self serve - and also have diesel fuel. The Owner is Kathy Davenport.

Bathrooms

The best advice is to go before you start on Highway 17.

There are virtually no public rest rooms anywhere along the length of the highway. Outside of the mountain areas, your best bet is to turn off the highway and look for a gas station or fast food restaurant. The services section of this book lists both in Los Gatos and Scotts Valley. In the mountain portion of the highway the only possibilities are the two restaurants at the summit and they discourage bathroom visits unless you are a customer.

During traffic jams, people have been known to pull off on a side road or go over the bank of the road looking for a private spot. While this is understandable, you should be aware that police can give you a ticket for urinating in public.

Other Services

Summit Properties located just south of Patchen Pass near the Summit Inn at 23111 Highway 17 (Los Gatos mailing address), phone (408) 353-1116. Owned by Bruce Kennedy, this business specializes in properties in the Summit area. One of his interests is classic American cars and he has a restored 1937 DeSoto housed in a building next to the real estate office.

Radiator Water

The best advice is to pray for rain. There is no place to get radiator water directly alongside Highway 17. Of course, the gas stations listed in the previous section have water, but most are not adjacent to the road.

The California State Automobile Association says that the most common emergency road call for Highway 17 is for overheating problems. Accidental breakages can happen to anyone, but many of the calls are simply a result of poor maintenance.

If your car loses all it's water, you had better call for a tow truck. If it is simply overheated, park in a safe place, turn off the motor and wait for it to cool down. That means at least half an hour. DO NOT REMOVE THE RADIATOR CAP. You can easily get burned and it does little to speed the cooling.

If you have air conditioning and your car is overheating, try turning the A/C off. It takes a great deal of power to run and often this alone will make a difference.

Hiking

There are many beautiful hiking areas in the Santa Cruz Mountains. Alongside Highway 17 there is the Henry Cowell Redwoods State Park; El Sereno, Saint Joseph's Hill and Sierra Azul open space preserves. If you are interested in more information, I highly recommend Tom Taber's *The Santa Cruz Mountains Trail Book*.

Accident and Density Patterns

Traffic Density Patterns

In 1983 there were 50,000 cars a day, in 1984 58,000, in 1986 74,000. Today an average of 75,000 vehicles use Highway 17 at the Lexington Reservoir point, every day, according to a 1990 study by David J. Powers and Associates for Caltrans. And as many as 160,000 vehicles pass at the 280 intersection daily.

By 1983 the Caltrans counts for Highway 17 had reached 3,400 per hour and the 1990 Santa Cruz Regional Transportation Commission report states that by 1995 there will be 4,100 southbound trips during the afternoon peak hour, and by 2005 it will reach 4,700 trips. The theoretical maximum used by traffic planners is 2,000 cars per hour per lane so the road is at it's maximum capacity already. Traffic will be considered at Caltrans level "F" during peak hours by 1995.

The busy commute hours are from 6 to 9 a.m., and from 4 to 7 p.m. 5-6 p.m. is peak traffic time. Highest traffic volumes occur during summer weekend times (9-11 a.m. southbound, 4-7 pm northbound). Traffic in Santa Cruz County increases 39% in the summer time due to visiting tourists. Fridays have 8% more traffic than average, Sundays 10% less. June is 8% higher than average, December 13% less.

The Highway 1-southbound Highway 17 intersection between 5 and 6 p.m. weekdays is the point of greatest density on a regular basis.

January 23, 1949 brings snow, and the police, to Highway 17. [UCSC *Special Collections*]

26,000 people commute from Santa Cruz County to Santa Clara County daily, and 6,700 people commute the opposite direction. In the next 10 years these figures will increase about 20%.

Trucks

There are no statistics available on the number of trucks using Highway 17 but commuters know there are a lot.

Some of the largest are the 65' sand trucks, weighting close to 80,000 pounds. The San Lorenzo Valley is probably the best sand deposit in the Western United States and there is enough sand to stay in business at least another 20 years.

Although large slow moving trucks are often a factor in accidents, they are rarely involved themselves in accidents with other cars. In general, truck accidents have been reduced over the past 10 years due to an extensive inspection program by the CHP. The more common occurrence for trucks are mechanical breakdowns or jackknifing due to excessive speed.

Many commuters believe that trucks are "hidden" causes of accidents as car drivers driving at the speed limit swerve and cut-in while attempting to avoid slow moving trucks. Periodically there are calls to ban trucks such as in 1988 when Arnold Wechter, the *Santa Cruz Sentinel* Automotive Editor called for banning large trucks during commute hours. However, the Federal Government, which funds much of Highway 17, does not allow any federally subsidized roads to ban trucks.

Truckers responded to complaints by voluntarily starting a "Truckers Right" program to request that trucks drive only in the right lane over the mountain sections of Highway 17 during commute hours.

In general, truck drivers are some of the most professional and safest drivers on the road. When interviewed, they uniformly worried about the driving styles of passenger car drivers on the road. As R & G Truck driver Brian Marshall said:

If I'm involved in an accident with a passenger car, I know I'll walk away safely because the truck is so massive. But the other driver is likely to be killed. It makes truckers very cautious.

Cutting in front of trucks is the biggest complaint - it takes a long time to stop 80,000 pounds! In general, it takes 3 times as much distance for a loaded truck to stop as it does for a passenger car, so cutting in front of trucks can be very dangerous to your health. And passenger cars slow unnecessarily as they pass trucks causing delays.

Accident Summary

Although Highway 17 is one of the most beautiful drives in California, most first time drivers come away complaining about the tight turns, poor visibility, lack of turnouts, poor camber and crazy drivers.

Despite all of the dangers, drivers should know that things are getting better. 1967 was the worst year on Highway 17 when 36 fatal accidents were recorded. In 1990 it was down to 8 fatalities. The concrete barrier gets most of the credit according to the CHP.

But with all of the safety improvements on an average there is an accident every day, 15 injuries a month, 4 people killed annually.

According to Caltrans, in 1990 there were 787 collisions on Highway 17. The primary collision factors, in order, were:

1. Speeding

2. Unsafe lane change

3. Improper turning (e.g., merging, left turns)

4. Driving under the influence

Highway 17 Accident Statistics

Year	Fatal Accidents	Injury Accidents	Property Damage
83	5	149	363
84	3	131	322
85	3	132	312
86	4	135	329
87	6	127	365

County	Accidents	Injuries	Fatalities
Santa Clara	547	293	5
Santa Cruz	240	145	3
Total	787	438	8

No. of Vehicles Involved In Accident

1	34%
2	49%
3	12%
More than 3	5%

Accident Direction of Travel

County	Northbound	Southbound
Santa Clara	61%	39%
Santa Cruz	41%	59%

Statistics from California Department of Transportation, 1990.

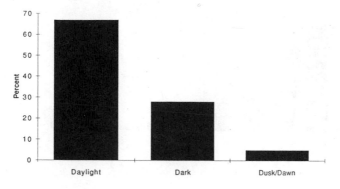

Accidents by Lighting Condition

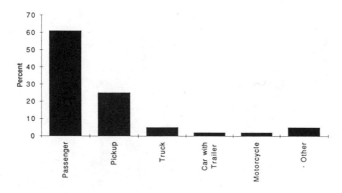

Accident by Vehicle Type

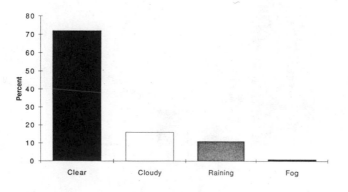

Accidents by Type of Weather

Statistics from California Department of Transportation, 1990.

Movement Preceeding Collision

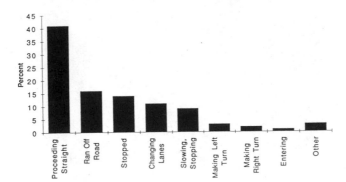

Type of Collision

Accidents by Road Surface

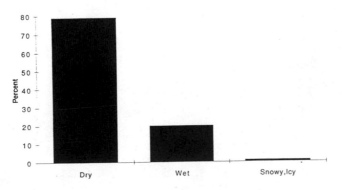

Statistics from California Department of Transportation, 1990.

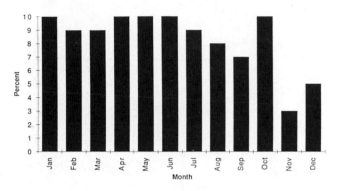

Accidents by Month

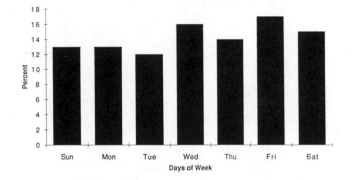

Accidents by Day of Week

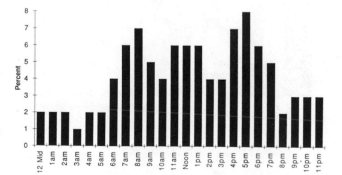

Accidents by Time of Day

Statistics from California Department of Transportation, 1990.

181

Primary Collision Factor

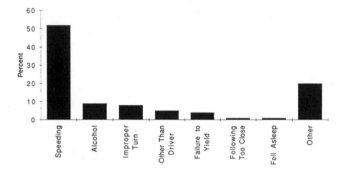

Objects Hit

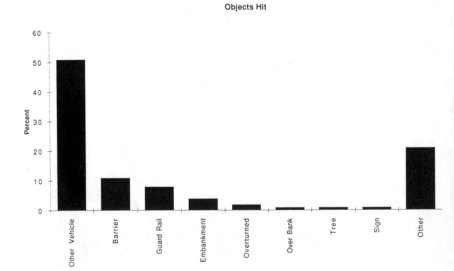

Statistics from California Department of Transportation, 1990.

APPENDIX 8

Emergency Response Services

If You See or Hear an Emergency Vehicle

If you see an emergency vehicle with it's emergency lights flashing, you are required by law to get out of the way.

Because of the narrowness of the hill portion of Highway 17, many emergency vehicle drivers prefer to go up the center of the road - between the two lanes. Look in your mirror, turn off your radio and listen for instructions on what to do.

You must stay 300 feet in back of emergency vehicles with their lights flashing. There is one exception. Going up hill, those 32,000 pound heavy fire trucks can't go much more than 25 m.p.h.. If they have only their amber lights flashing and are in the right hand lane, you may pass on the left, SLOWLY and CAREFULLY. If the red lights are flashing, you cannot pass. And if you pass an emergency vehicle, realize there is some kind of an accident scene up ahead. Be extra careful.

California Highway Patrol

The California Highway Patrol is primarily responsible for police protection and response on Highway 17. Two separate offices handle the responsibility - based on the two county boundaries. However, CHP officers can cross their beat

Alma Fire Station before it was relocated due to the Lexington Reservoir construction.
[San Jose Historical Museum]

boundaries and can issue citations on any road at any time. Normally the offices divide responsibility at the summit county line, but often respond to a call in each others area if they have an officer nearby - until the normal office can respond. Both patrol routes use the Summit Garden Inn restaurant at the summit as the place to stop for coffee and meal breaks.

The Santa Clara County office is located at 2020 Junction Avenue, San Jose (408) 277-1800. That's near the Brokaw exit of Highway 17. The office is responsible for all state highways within Santa Clara County plus several other small areas and is considered part of the "Golden Gate" CHP Region. 128 Officers are assigned to the office. Santa Clara officers are dispatched from a new center in Vallejo.

The Santa Cruz County office is located at 10395 Soquel Drive in Aptos (near the Freedom Blvd. exit off Highway 1) and is part of the "Coastal" CHP Region. The administrative phone is (408) 662-0511. Santa Cruz County CHP officers are dispatched through a dispatcher located in Monterey. The dispatcher's non-emergency phone number is (408) 455-1823.

Local police departments can pursue violators on Highway 17 and can only issue tickets for violations occurring within their jurisdiction. Be aware that some sections of Highway 17 are within city limits (e.g., Scotts Valley, Los Gatos, Campbell, San Jose). This is rarely done, however, and for all practical purposes the CHP is primary police force.

Santa Cruz normally has one officer assigned to their side of 17 and Santa Clara has two officers on patrol - although in times of higher traffic or when a speed monitoring program in effect, there can be as many as 7 CHP officers assigned to each side of Highway 17.

The CHP has four types of specially modified vehicles currently in use:

- Chevrolet police cruisers

- Ford Mustang highway pursuit vehicles

- Specially marked cars or pickup trucks used to inspect commercial vehicles. (a commercial vehicle has 3 axles, weights 26000 pounds or more, or is used totransport people).

- Motorcycles

Both CHP departments have used radar regularly on the expressway sections since 1986 to check on vehicle speeds. The CHP is prohibited by state law from using radar on freeways (but local law enforcement agencies, e.g., Scotts Valley Police, can use radar within their city limits).

Fire Departments

Campbell Fire Department

The Campbell Fire District has responsibility for emergency responses within their city limits, but do cooperative aid north to Highway 80 and south to Lark Avenue.

They have 36 paid fire and emergency medical personnel, assisted by 20-25 volunteers. Most of their Highway 17 calls are medical, averaging 10-12 a month. Responses are made by a single fire engine with a paramedic. A second engine with a Batallion Chief will response if more help is needed. Campbell has their own dispatching services where 911 and call box calls are routed to that center.

The main Campbell fire station is located at 123 S. Union in Campbell, with a second station is at 485 W. Sunny Oaks. The administrative phone number is (408) 866-2189.

Scotts Valley Fire District

The district is responsible for fire and rescue operations on Highway 17 from the Santa Cruz City Limit just north of Highway 1 and the Spanish Oaks Rancho Road which is about one-third mile north of Laurel Curve.

There are 22 paid fire personnel and 14 volunteers. Notification comes through the Santa Cruz County 911 dispatcher. Fire/rescue personnel are dispatched to all accident scenes, along with CHP and an ambulance unless there is prior information that assures they will not be needed. Response to Highway 17 calls are from 3 to 10 minutes depending upon the location. The Scotts Valley Fire Department dispatches a full fire engine with between 3 and 5 personnel to the accident scene. Each fire engine contains full rescue gear plus 500 gallons of water for fire fighting. While they have foam capabilities on the truck, water is usually much more effective. Injured victims are normally taken to Dominican Hospital in Santa Cruz.

They receive about 150 calls a year to Highway 17 accidents, with 95% of the calls being injury related. The biggest problem for rescue personnel are the "over the edge" accidents where the car leaves the road and falls down the embankment.

The department headquarters is at 7 Erba Lane, Scotts Valley , (408) 438-0211 for business calls.

Santa Cruz County Fire Department

This department is responsible for fire and rescue operations in the unincorporated areas of Santa Cruz County, including Highway 17 from the Summit south. They have primary responsibility from the Summit to Valley Oaks Road, and mutual aid responsibility with Scotts Valley Fire Department from there south to the Glenwood cutoff.

This Santa Cruz County section of the Highway is covered by the California Department of Forestry and Fire Protection's (CDF) Burrell Station (25050 Highland Way, 408-335-1022) and the Loma Prieta Volunteer Fire Station on Old Summit Road, about a quarter mile off the Highway. Together they get about 100 calls a year to the highway. CDF normally have five three fighters on duty, and there are another 30 volunteers in the area, about half who have EMT personnel, and the response depends upon the exact nature of the reported problem. Medical calls also have helicopter response available for emergency injury transportation. The County Fire Department gets calls relayed from the 911 center in Santa Cruz, and uses the CDF dispatching center in Felton.

Incidentally, the Alma Fire Station does not have primary responsibility to respond to incidents on Highway 17. The Forestry Department's primary mission is protection of brush in watershed areas. They do respond to incidents on a mutual aid basis when other units are not available but on an infrequent basis. The Alma station is only open during the fire season.

Central Fire District

The area northbound from the summit to Camden, and southbound from Lark Avenue to the summit is the primary responsibility of the Central Fire District headquartered in Los Gatos. Four stations primarily respond to incidents:

- Winchester Station - Winchester and Lark

- Los Gatos Station - on University Avenue

- Shannon Station - Shannon Road and Cherry Blossom Rd. in Los Gatos

- Redwood Station - located in Redwood Estates

All of their personnel are EMT trained, plus most have additional "D" certification that enables them to use the heart defibrillator. Normally one engine with 3-4 fire personnel responds to an incident, more if there are known injuries or a rescue. Injured people are taken to the Valley Medical Center Trauma Center,

751 S. Bascom, San Jose - take the Hamilton Exit - (408) 299-5100. For minor scrapes or injuries people can also be taken to Los Gatos Community or Good Samaritan.

The Central Fire District responds to 150-200 calls a year on Highway 17, 50 of which are serious incidents. Response time is normally five minutes or less. They recently upgraded their fire trucks with more powerful engines so they can move at the speed limit up the hill. Central County Fire is dispatched through the CHP dispatch center or Santa Clara County Communications depending on whether you use 911 or cellular phones.

Their headquarters is located at 14700 Winchester Blvd., Los Gatos and their administrative number is (480) 378-4010. Ben Lopes is the Deputy Chief of Operations.

Tow Trucks

There are many tow truck companies involved with Highway 17. If you belong to the AAA, Mobil, etc. type of service, police and dispatchers will call them for you. If you don't have a tow truck service, or your car is being impounded, officials simply call on a rotating basis from a list they have of authorized companies.

If your car is impounded, you pay all the costs. If your car has broken down and you simply need towing, most tow truck drivers require payment in cash, a local check or many accept credit cards. Be sure to find out the cost before you agree to anything.

Ambulance

The fire departments involved rely upon private ambulance companies to provide paramedics and ambulances. The list of approved companies is continually changing, and companies are called on a rotating basis. If an ambulance is called for you and you are taken to a hospital, the ambulance company will bill you for their charges.

Camps Garage 1910 tow truck (phone number Los Gatos 82). *[William A. Wulf]*

Los Gatos Main Street automobile repair business about 1907. *[Los Gatos Library]*

APPENDIX 9

Detail Maps and Listings

Introduction

Following are detailed maps and listings about the significant items along Highway 17.

There are two detailed maps:

- Santa Clara County
- Santa Cruz County

and detail listings showing

- Mileage marker signs
- Intersecting roads and driveways
- Roadway signs (speed limits, city limits, etc.)
- Locations of public telephones
- Locations of emergency call boxes
- Businesses along the way

The two listings in the Appendix first detail the highway as it appears driving South to North, then North to South.

Map of
Santa Clara County

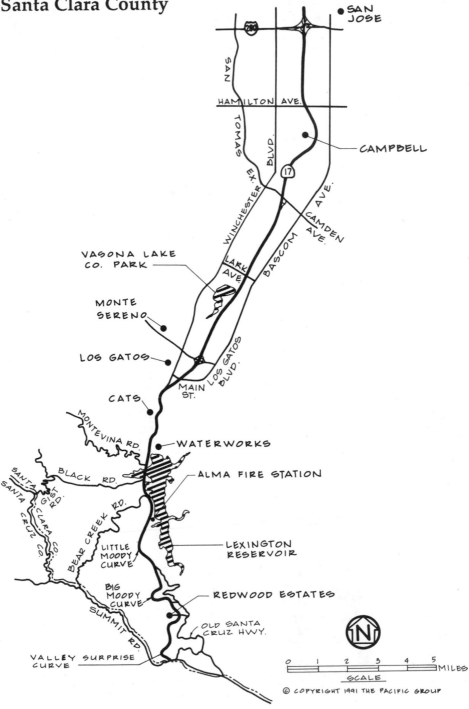

Map of
Santa Cruz County

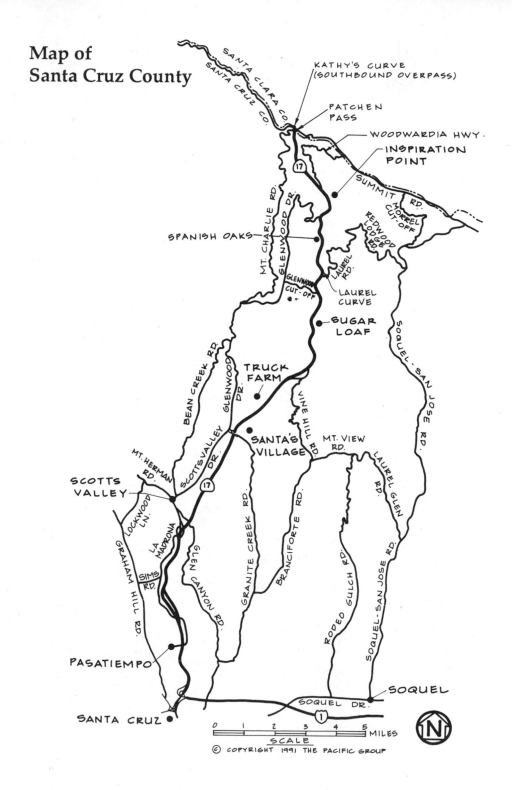

KATHY'S CURVE
(SOUTHBOUND OVERPASS)

PATCHEN
PASS

WOODWARDIA HWY.

INSPIRATION
POINT

SANTA CLARA CO.

SANTA CRUZ CO.

17

SUMMIT RD.

MORREL CUT-OFF

SPANISH OAKS

MT. CHARLIE RD.

GLENWOOD DR.

REDWOOD LODGE RD.

LAUREL RD.

GLENWOOD CUT-OFF

LAUREL CURVE

SUGAR LOAF

SOQUEL-SAN JOSE RD.

BEAN CREEK RD.

GLENWOOD DR.

TRUCK FARM

VINE HILL RD.

SANTA'S VILLAGE

MT. VIEW RD.

SCOTTS VALLEY DR.

MT. HERMAN RD.

SCOTTS VALLEY

17

LOCKWOOD LN.

LA MADRONA

SIMS RD.

GLEN CANYON RD.

GRANITE CREEK RD.

BRANCIFORTE RD.

LAUREL GLEN RD.

RODEO GULCH RD.

SOQUEL-SAN JOSE RD.

GRAHAM HILL RD.

PASATIEMPO

SOQUEL

SOQUEL DR.

1

SANTA CRUZ

0 1 2 3 4 5 MILES
SCALE

N

© COPYRIGHT 1991 THE PACIFIC GROUP

Detail Listing from South to North

Highway 1 Interchange
00.26
00.29
00.35
00.43
00.50
Pasatiempo Drive (right to El Rancho Dr. & La Madrona Dr, left to Pasatiempo)
o/h bridge Pasatiempo Road
00.79
00.87
01.00
01.10
01.22
Call Box SZ-017-014
01.44
01.50
01.62
01.76
01.94
End Freeway Sign
02.03
02.15
Sims Road & Graham Hill Road (right to El Rancho Dr. & La Madrona Dr.)
02.31
02.40
Begin Freeway Sign
02.48
02.50
02.62
02.83
02.97
Scotts Valley City Limits
03.23
Mount Hermon Road (to Big Basin, Felton and Scotts Valley)
03.43
o/h bridge - Mount Hermon Road
03.50
03.59
03.81
Call Box SZ-017-038
04.00
04.12
04.27
04.50
04.52
Call Box SZ-017-046
04.90
04.97
05.21
Call Box SZ-017-052

05.37
05.48
Granite Creek Road and Glenwood Drive
o/h bridge - Granite Creek
05.56
05.71
End Freeway sign
Santa's Village Road
05.91
Daylight Safety Section sign
Call Box SZ-017-062
Speed Limit 50 sign
06.07
private driveway
06.23
turnout
06.33
06.43
Errington Lane (private)
06.50
private driveway
06.53
Crescent Drive
Crescent Drive (leads to Crescent Court)
06.72
06.80
left Truck Farm
Call Box SZ-017-068
06.86
West Vinehill Road (turn right to Vinehill Road or left loops back to 17)
07.15
07.23
07.33
Vinehill Road (keep left or right loops back to 17)
Jarvis Road
private driveway (loops to next)
private driveway (loops to previous)
07.46
Speed limit 50 sign
07.50
Call Box SZ-017-076
07.73
07.88
turnout
08.16
driveway
Call Box SZ-017-082
08.21
08.28
08.34
08.40
08.51
Speed limit 50 sign

driveway
08.63
turnout - "sandpile" and Sugarloaf Road
Phone
08.73
Carl Drive
08.94
left Glenwood Cutoff
driveway
09.15
09.33
turnout
09.41
Call Box SZ-017-094
Laurel Road leading to Laurel (left) and old Santa Cruz Highway (right)
unpaved driveway
turnoff
turnoff
09.47
driveway
09.60
09.85
Call Box SZ-017-098
Speed limit 50 sign
road
09.99
driveway
10,00
10.10
driveway
10.30
10.44
turnoff
Call Box SZ-017-106
Phone
10.52
10.60
left Glenwood Road
10.82
Inspiration Point turnout
Call Box SZ-017-108
Phone
10.91
11.00
driveway
driveway
11.12
Call Box SZ-017-114
11.34
Old Santa Cruz Highway
11.50
left Summit Inn (phone)
Cloud 9 (phone)

Speed limit 50 sign
11.96
turnout
12.52
Patchen Pass
Santa Clara County limits
Summit Road (phone)
Wreck sign
overpass - summit road
Call Box SC-17-02
Truck speed limit 35 sign
00.15
Speed Limit 50 sign
Call Box SC-17-06
turnout
Truck speed limit 35 sign
00.75
01.00
Call Box SC-17-12
01.11
Madrone Rd. (Redwood Estates & Holy City - leads to old Santa Cruz Highway)
Speed limit 50 sign
01.40
01.77
Truck speed limit 35 sign
Call Box SC-17-16
01.65
01.75
01.80
01.81
02.00
Call Box SC-17-22
02.20
Rainer Lane
Speed limit 50 sign
02.30
02.35
02.50
Idylwild Road (leads to old Santa Cruz Highway)
left road
Hebard Road
turnoff
Call Box SC-17-28
Phone
Truck speed limit 35 sign
03.00
Speed limit 50 sign
Call Box SC-17-34
03.35
Old Santa Cruz Highway
03.45
03.50
03.60

Alma Fire Station
Phone
03.70
Truck speed limit 35 sign
Call Box SC-17-36
03.99
Lexington Reservoir
04.00
left Bear Creek Road
Call Box SC-17-42
Speed limit 50 sign
04.18
04.22
left Black Road
turnoff
04.50
04.60
left Montevina Road
Alma Bridge Road
Speed Limit 50 sign
Waterworks Station
Truck 35 m.p.h. sign
End daylight safety test sign
05.00
Call Box SC-17-52
05.05
06.70
Call Box SC-17-56
05.77
Start Freeway sign
55 m.p.h. sign
End truck speed limit 35 sign
05.85
06.15
left The Cats
Begin freeway sign
Speed limit 55 sign
06.15
North Santa Cruz Avenue (exits from left lane)
06.25
06.30
06.35
o/h Main Street bridge
06.65
o/h pedestrian overcrossing
06.70
Call Box SC-17-66
East Los Gatos exit
07.00
Highway 9 - Saratoga Road (to Monte Serano)
o/h bridge Saratoga Road
07.40
Call Box SC-17-76

o/h Blossom Hill Road bridge
07.75
07.80
07.85
07.95
08.00
08.10
08.20
Call Box SC-17-82
Call Box SC-17-86
Lark Avenue
08.75
o/h bridge Lark Avenue
08.75
Vasona Park on the left
08.80
bridge
08.90
Call Box SC-17-88
09.00
09.02
09.23
Call Box SC-17-94
Campbell City Limit
09.60
09.70
Los Gatos County Park to the west
Call Box SC-17-98
Camden Avenue & San Tomas Expressway
Call Box SC-17-106
10.47
10.55
3 lanes
10.80
11.45
Pruneyard
12.01
Hamilton Avenue
o/h train overcrossing
12.31
o/h bridge Hamilton
12.35
Call Box SC-17-124
4 lanes
Call Box SC-17-128
San Jose City Limit
Call Box SC-17-134
13.50
13.65
13.68
13.72
Highway 280

Detail Listing From North To South

Highway 280
5 lanes
Call Box SC-17-135
pedestrian overcrossing
Campbell City Limits
Call Box SC-17-129
Hamilton Avenue
3 lanes
Call Box SC-17-125
o/h bridge (Hamilton)
12.30
12.20
o/h bridge (train)
3 lanes
Pruneyard
Camden Avenue & San Tomas Expressway
2 lanes
10.65
Call Box SC-17-107
o/h bridge (Camden)
10.53
10.30
10.15
3 lanes
10.00
Call Box SC-17-99
09.90
09.85
09.75
Los Gatos City Limits
Call Box SC-17-95
09.00
Lark Avenue
2 lanes
Call Box SC-17-89
o/h bridge (Lark Avenue)
08.90
08.85
08.50
Call Box SC-17-87
08.10
Call Box SC-17-83
Vasona Park
07.85
Speed Limit 55 sign
07.70
0/h bridge (Blossom Hill Road)
Call Box SC-17-77
07.50
07.40
07.15

Highway 9 - Saratoga Road (Los Gatos, Saratoga)
East Los Gatos
07.00
Call Box SC-17-67
pedestrian overcrossing
06.60
o/h bridge (Main Street)
06.45
06.35
06.15
06.05
Speed Limit 50 m.p.h. sign
end freeway sign
05.90
The Cats
05.75
Call Box SC-17-57
5.60
driveway
05.50
05.30
05.27
05.15
turnout
Call Box SC-17-53
Speed Limit 50 sign
05.09
05.00
Daylight Safety Section
04.97
04.95
left San Jose Water Works
04.81
left Alma Bridge Road
Montevina Road (phone)
Speed Limit 50 sign
Lexington Reservoir
04.64
04.45
Black Road
04.40
04.35
Call Box SC-17-47
04.18
Bear Creek Road (leads to Summit Road)
04.00
03.95
03.85
03.75
Speed Limit 50 sign
left Alma Fire Station (phone)
03.50
left Old Santa Cruz Highway

turnout
Call Box SC-17-35
Phone
Alma College Rd
03.40
turnout
03.10
Call Box SC-17-33
03.01
03.33
02.99
02.94
02.80
Hebard Rd
02.75
Speed limit 50 sign
turnout
Call Box SC-17-27
Phone
left Idylwild
02.55
02.50
02.45
Hillside Drive
02.40
02.35
Brush Road
02.00
Call Box SC-17-19
01.80
turnout
01.78
Speed Limit 50 sign
01.40
01.38
01.36
01.30
Redwood Estates (to Holy City, Old Santa Cruz Highway)
(phone)
00.95
Call Box SCL-017-009
Speed Limit 50 sign
00.77
00.65
turnout
00.15
Summit Road (phone)
o/h bridge (Summit Rd.)
Patchen Pass
12.55
Santa Cruz County Line
12.52
Speed limit 50 sign

turnout
Phone
Call Box SZ-017-125
12.34
12.28
turnout
12.20
Call Box SZ-017-123
Speed Limit 50 sign
Truck 35 m.p.h. sign
left - Cloud 9
11.78
Summit Inn (phone)
11.50
turnout
driveway
11.34
left - Old Santa Cruz Highway
turnout
driveway
11.26
11.03
turnout
Call Box SZ-017-113
Phone
11.00
10.91
Glenwood Drive
10.56
turnout
Call Box SZ-017-105
10.40
driveway
10.36
10.29
Spanish Oaks driveway
turnout
Call Box SZ-017-103
10.00
Speed limit 50 sign
09.85
Laurel Curve
turnout
Call Box SZ-017-097
09.60
09.54
09.50
09.47
left - Laurel
turnout
Call Box SZ-017-095
Truck speed 35 sign
09.22

09.15
Glenwood Cutoff (leads to Glenwood Highway)
09.00
turnout
driveway
Call Box SZ-017-089
08.89
08.77
08.73
left - "sand pile" and Sugarloaf Road
08.63
driveway
Call Box SZ-017-087
Speed limit 50 sign
08.51
Truck speed 35 sign
08.40
08.34
08.28
08.21
turnout
07.99
turnout
Call Box SZ-017-079
07.88
07.73
Speed limit 50 sign
07.57
07.46
left - Vinehill
driveway
07.45
Speed limit 50 sign
07.23
07.15
Scotts Valley City Limits
07.07
turnout
Call Box SZ-017-073
07.00
turnout
Truck Farm
Call Box SZ-017-069
driveway
06.86
06.80
left - driveway
driveway
driveway
DeForges
06.53
06.50
turnout

Truck speed 35 sign
06.33
06.23
Begin Freeway sign
06.07
End truck 35 limit sign
end 50 m.p.h. sign
End daylight section
Call Box SZ-017-059
05.71
Granite Creek Road (Scotts Valley, Bethany College)
05.50
o/h bridge (Granite Creek)
05.37
Call Box SZ-017-053
05.00
04.97
Call Box SZ-017-047
04.52
04.50
04.43
04.40
04.27
04.00
03.56
Call Box SZ-017-039
Glenwood Rd & Mount Hermon Road
03.10
02.50
end freeway sign
02.31
02.20
Sims Road & Graham Hill Rd
02.30
begin freeway sign
01.94
01.84
01.62
Call Box SZ-017-015
01.22
01.10
01.00
Pasatiempo Drive
00.79
overhead bridge - Pasatiempo
00.50
00.43
00.20
Highway 1 Interchange

Reference Materials

Most of the information in this book was obtained from the following authors and historians who have my admiration for their outstanding original research. After spending hundreds of hours in dusty library files, I can only begin to appreciate how hard it must have been for those who went first. My thanks to you all!

Anonymous. "Charles H. Purcell is Dead," *Western Construction* magazine, October 1951, p. 99.

Ayres, Harriet Springman. *My Memories of Santa Cruz and Scotts Valley*. Unpublished paper, September 1986. *A copy can be viewed at the Santa Cruz County Historical Trust in Scotts Valley.*

Barriga, Joan B. *Survival with Style: The Women of the Santa Cruz Mountains.* Unpublished paper to the Californians, Westward Ho!. 1989 (rev). *A copy can be viewed at the Los Gatos Public Library.*

Barriga, Joan B. *The Holy City Sideshow.* Unpublished paper to the California Pioneers of Santa Clara County Essay Contest. April 20, 1988. *A copy can be viewed at the Los Gatos Public Library.*

Barriga, Joan B. *The Raucous Bluejay: Charles Erskine Scott Wood.* Unpublished paper to the California Pioneers of Santa Clara County, April 21, 1989. *A copy can be viewed at the Los Gatos Public Library.*

Beilharz, Edwin A. and Donald O. Demers. *San Jose: California's First City.* Continental Heritage Press, Oklahoma, 1980.

Bergstrom, Mark. "Highway 17 is the Mata Hari of freeways" *Santa Cruz Sentinel,* March 26, 1987, pE4.

Bernstein, Bob. *Peninsula Living,* May 8, 1954, p. 9.

Bruntz, Dr. George G. *The History of Los Gatos: Gem of the Foothills.* Valley Publishers, Fresno, Ca. 1971.

California Department of Highways files. Road Construction Files, 5A, 1914-1922. Photographic albums 1912-1949. State Archives office, Sacramento. *Unfortunately very few of the original documents have survived.*

California Division of Highways. *Final Report for the Construction of a Portion of the State Highway in Santa Cruz County From One Mile North of Inspiration Point to Scotts Valley,* November 10, 1934, contract 64TC17-44E09-24EC1, IV-SCr-5-B.

California Highways Commission. *California Highways and Public Works,* October 1934; October 1935; November 1937; October 1938; May 1940; September 1940; December 1941; September 9, 1950; August 5, 1970. Published by the California State Highway Commission, Sacramento.

Cabinland Magazine, published by the Redwood Estates Development Company, San Francisco. February and June 1927. *Available at the Los Gatos Public Library.*

Clark, Donald Thomas. *Santa Cruz County Place Names*. Santa Cruz Historical Society, 1986. *An excellent reference book.* (Clark is UC Librarian Emeritus.)

Francis, Phil. *Beautiful Santa Cruz County, 1896. A copy is at the University of California Santa Cruz Special Collections Library.*

From Shoreline To Skyline, Roadrunner Audio Tour. Roadrunner Audio, Inc., Albuquerque, NM., 1988.

Garaventa, Donna M. and Robert M. Harmon. *Historic Property Survey Report for Route 17 at Lexington Reservoir Interchange and Frontage Road Project, Santa Clara County, CA* Basin Research Associates, Inc., 14731 Catalina St., San Leandro, Ca. November 1989, Revised April 1990.

Gilroy Dispatch. August 28, 1988.

Hamman, Rick. *California Central Coast Railways*. Pruett Publishing Company, Boulder, Colorado, 1980.

Hornbeck, David. *California Patterns*: A geographical and historical atlas. Mayfield Publishing Company, 1983.

Kenneally, Finbar, translator and editor. *Writings of Fermin Francisco De Lasuen*, vol.1, p. 235-7, Academy of American Franciscan History, Washington, D.C., 1965.

Koch, Margaret. *Santa Cruz County, Parade of the Past*. Valley Publishers, Fresno, CA, 1973.

Los Gatos Times Observer, January 7, 1978.

Lydon, Sandy. *Chinese Gold: The Chinese in the Monterey Bay Region*. Capitola Book Company, Capitola, CA, 1985.

Michel, Peter. *17*, a research paper prepared for a University of California Santa Cruz class in 1988. *Held at the UCSC Special Collections Library. Portions published that same year in the Santa Cruzian newspaper.*

Olin, L.G. *The Development and Promotion of Santa Cruz Tourism*, University of California Santa Cruz thesis, 1967. *An excellent paper with many interesting facts.*

Patton, Phil. *Open Road, A Celebration of the American Highway*. Simon and Schuster, New York.

Payne, Stephen. *A Howling Wilderness, The Summit Road of the Santa Cruz Mountains 1850-1906*. Loma Prieta Publishing, Los Gatos, CA, 1978. *A classic and fascinating book.*

Payne, Stephen M.. *Santa Clara County: Harvest of Change*. Windsor Publications, Northridge, CA, 1987.

Petersen, Grant and John Kluge. *Roads to Ride, South*. Heyday Books, Berkeley., 1985.

Primack, Mark. *Axel Erlanson's World Famous Tree Circus.* Originally published in consecutive issues of Zeitgeist. *A copy can be viewed at the UCSC Special Collections Library.*

Regional Transportation Plan, Santa Cruz County. Santa Cruz County Regional Transportation Commission, August 1990.

Rowland, Leon. *Santa Cruz: The Early Years,* Paper Press, Santa Cruz, CA, 1980.

Sawyer, Preston. Santa Cruz newspaper (no title, Sentinel?), August 31, 1921.

San Francisco Chronicle, February 21, 1985, p.7.

San Jose Mercury, December 5, 1953; January 8; 1954; September 21, 1954; July 26, 1956; September 16, 1957; May 26, 1966; October 3, 1968; August 27, 1969; February 15, 1971; March 9, 1971; June 26, 1974; July 25, 1974; July 26, 1974; October 10, 1974; November 16, 1974; August 27, 1978, page 18A; June 17, 1984; June 10, 1985; June 17, 1985; September 23, 1987, August 9, 1989. *Other clippings were examined on the "highway" microfiche files at the San Jose Mercury , but in many cases there were no dates recorded.*

Santa Cruz Sentinel, December 1, 1915; September 5, 1935; December 19, 1937; December 17, 1950; May 8, 1954; January 16, 1955; December 16, 1966; January 12, 1967; February 28, 1971; July 6, 1978; March 26, 1987; February 10, 1988; December 4, 1988; April 29, 1991. *Many of the citations used in this book came from articles found the Sentinel's clipping files, which unfortunately did not include page numbers.*

Taber, Tom. *The Santa Cruz Mountains Trail Book.* The Oak Valley Press, San Mateo, 6th edition, 1991. *Excellent book.*

Thomas, John Hunter. *Flora of the Santa Cruz Mountains of California.* Stanford University Press, Palo Alto, California, 1961.

Transportation 2000: Final Plan May 1987. Santa Clara County Transportation Agency, San Jose, California.

T2010 Santa Clara County Transportation Plan (Draft Final Plan) July 1991. Santa Clara County Transportation Agency, Advance Planning Unit. San Jose, California.

Young, John V. *Ghost Towns of the Santa Cruz Mountains.* Western Tanager, Santa Cruz, Ca, 1984. *Much of the material forming this book was originally published in 1934 in the San Jose Mercury Herald as part of Young's Sunday feature series.*

U.S. Army Map Service,. *Ben Lomond quadrangle map , 1940, 1943 and 1966 , 1/ 50,000 and 1/62,500 scales.*

U.S. Department of the Interior, U.S. Geological Survey. *Los Gatos quadrangle map (mislabelled New Almaden), 1919, 1/62,500 scale.*

U.S. War Department, Corp of Engineers, U.S. Army. *Los Gatos quadrangle map, 1940 and 1943, 1/62,500 scale.*

Wallace, Janette Howard. *Reminiscences of Janette Howard Wallace.* Unpublished paper, 1986. *A copy can be viewed at the Santa Cruz County Historical Trust in Scotts Valley.*

Watkins, Major Rolin C. *History of Monterey, Santa Cruz and San Benito Counties, California*, S.J. Clarke Publishing, Chicago, 1925.

Watson, Jeanette. *Campbell, the Orchard City.* Campbell Historical Museum and Association, Campbell, California. 1989.

Wulf, William A. *A History of the Santa Cruz Mountains and Early Los Gatos.* Unpublished paper, 1990. *A copy can be viewed at the Los Gatos Library Reference Room.*

Interviews and Other Assistance

This book would not have happened without invaluable assistance from the people listed below.

Jim and Peggy Alberti	MasCom, cover design
Harriett Ayres	Early resident of Scotts Valley, near Sims Road and Highway 17 area
Gregory Bayol	California Department of Transportation, District IV, Public Information Officer
Marlyn Bergman	Scotts Valley Realty, Owner
Irene Berry	University of California, Santa Cruz Special Collections Library
Horace Bristol	Assisted in obtaining photograph of Charles Erskin Scott Wood
Scotty Bruce	Santa Clara County Transportation Agency, Deputy Director Design & Construction Division
Chief O.J. Burrell	California Dept. of Forestry and Fire Protection - Region 1
Lisa Christenson	California History Center, DeAnza College
Laurel Clark	California Department of Transportation, Sacramento Archives Librarian
Ed Costa	California Department of Transportation District IV, Maintenance Manager

Lewis Deasy	Santa Cruz Historian, specializes in Capitola
Marc Eymard	County of Santa Cruz, Planning Department Supervising Planner
Barry Evans	Publisher, Garden Court Press
Rick Hamman	Eccles & Eastern Railroad Company, President
Jack Hanson	Eccles & Eastern Railroad Company, Vice President
John Hesler	David J. Powers & Associates, Environmental Consultants & Planners
Bruce Hoffman	City of Scotts Valley, Public Works Department
Rochelle Hooper	California Department of Transportation, Photo Lab, Sacramento
Hal Hyde	Senior Vice President, Ford's Department Store, long time Santa Cruz County family
Sister Mary Joseph	St. Clare's Retreat House, Missionary Sisters of Our Lady of Sorrows
Bruce Kennedy	Owner of Summit Properties and area Historian
Stephen King	G.M. Drafting
Jean Kirtchell	Los Gatos Library Special Collections
Margaret Koch	Longtime resident of Scotts Valley area, well known Writer and Historian
Gary Lance	San Jose Mercury Research & Library Department
Betty Lewis	Area Writer and Historian, specializes in south Santa Cruz County but has a new book on Holy City in the works
Ben Lopes	Santa Clara County Central Fire District, Deputy Chief of Operations
Sandy Lydon	Santa Cruz Historian, teacher at Cabrillo College
Brian Marshall	R & G Trucking, sand truck driver
Leslie Masunaga	Archivist, San Jose Historical Museum
Rachel McKay	Santa Cruz Historical Museum, archivist

Chief Mike McMurray	Scotts Valley Fire Department
Stephen Payne	Owner Loma Prieta Properties, San Jose State teacher, area historian and writer
Marion Pokriots	Scotts Valley Historical Society
Karena Pushnik	County of Santa Cruz Transportation Department
James Ross	California Department of Transportation, District IV, District Hazardous Waste Coordinator, Aerial Photo Files
David Schneider	California State Archives, Sacramento
Officer Mary Schuldes	California Highway Patrol, Public Information Officer, San Jose
Chief Bruce Scott	Scotts Valley Fire Department Department Chief
Martin Shore	Summit resident, great grandson of Charles C. Martin
Nikki Silva	Curator, Santa Cruz Historical Trust
Chief Steve Staump	Batallion Chief, Campbell Fire Dept.
Stanley D. Stevens	University of California, Santa Cruz McHenry Library Map Room
Glyn Stout	Lupin Naturist Club, Owner
Ruby V. Strong	Early Scotts Valley resident
Paul Stubbs	University of California, Santa Cruz Special Collections Library
Nancy Valby	Curator, San Jose Historical Museum
Steve Wilson, KA6S	Santa Clara Ham Emergency Services Coordinator.
Linda Wilshusen	Santa Cruz County Transportation, Executive Director
Officer Ernie Winsor	California Highway Patrol, Santa Cruz County
William A. Wulf	Historian and Writer, specializing in early history of the local area and Los Gatos

About the Author

Lighthouse Photography

Eleventh generation American and native Californian Richard Beal, 48, is a widely known management consultant in the computer industry where he has worked with companies such as Apple Computer, Digital Equipment Corporation, James River Corporation, NEC Electronics, Bechtel and Pacific Bell to design advanced computer applications. He and wife Kathy own a consulting firm called The Pacific Group, with offices in San Francisco and Aptos, California. Free time is spent traveling, visiting art galleries, attending the ballet and symphony or watching sport car races. Richard has a BS degree from UCLA in Finance, an MS in Sociology from the University of Missouri and a Masters of Divinity degree in Theology from San Francisco Theological Seminary. During the mid-60s Richard served as a Captain in the US Air Force. He has two grown children: son Andrew (21) who is studying at Cabrillo College, and daughter Chandra (24) who is at the University of Texas, Austin.

How To Order This Book

Single or multiple copies of this book can be ordered at any bookstore, or directly from the publisher at:

The Pacific Group
PO Box 44
Aptos, California 95001
FAX (408) 662-0934

The single copy price direct from the publisher is $12.95 plus your local sales tax plus $3.00 for shipping. Visa and Mastercard are accepted. Please include the card number, expiration date and the name as it appears on the card, plus a daytime phone number in case there are questions.